Sophia Studies in Cross-cultural Philosophy of Traditions and Cultures

Volume 47

The Sophia Studies in Cross-cultural Philosophy of Traditions and Cultures focuses on the broader aspects of philosophy and traditional intellectual patterns of religion and cultures. The series encompasses global traditions, and critical treatments that draw from cognate disciplines, inclusive of feminist, postmodern, and postcolonial approaches. By global traditions we mean religions and cultures that go from Asia to the Middle East to Africa and the Americas, including indigenous traditions in places such as Oceania. Of course this does not leave out good and suitable work in Western traditions where the analytical or conceptual treatment engages Continental (European) or Cross-cultural traditions in addition to the Judeo-Christian tradition. The book series invites innovative scholarship that takes up newer challenges and makes original contributions to the field of knowledge in areas that have hitherto not received such dedicated treatment. For example, rather than rehearsing the same old Ontological Argument in the conventional way, the series would be interested in innovative ways of conceiving the erstwhile concerns while also bringing new sets of questions and responses, methodologically also from more imaginative and critical sources of thinking. Work going on in the forefront of the frontiers of science and religion beaconing a well-nuanced philosophical response that may even extend its boundaries beyond the confines of this debate in the West – e.g. from the perspective of the 'Third World' and the impact of this interface (or clash) on other cultures, their economy, sociality, and ecological challenges facing them – will be highly valued by readers of this series. All books to be published in this Series will be fully peer-reviewed before final acceptance.

Michael J. Ardoline

Interiority Against Control

The Metaphysics, Ethics, and Politics of Information Technologies through Deleuze and Confucianism

Springer

Michael J. Ardoline
Louisiana State University
Baton Rouge, LA, USA

ISSN 2211-1107 ISSN 2211-1115 (electronic)
Sophia Studies in Cross-cultural Philosophy of Traditions and Cultures
ISBN 978-3-032-16244-1 ISBN 978-3-032-16245-8 (eBook)
https://doi.org/10.1007/978-3-032-16245-8

This Springer imprint is published by the registered company Springer Nature Switzerland AG
The registered company address is: Gewerbestrasse 11, 6330 Cham, Switzerland

Acknowledgments

This is a book I did not intend to write. Rather, it was already half complete by time I realized it was a book at all. Many of its pieces came about first because I was asked by friends and colleagues to write or present on various topics that I initially took to be unrelated. It was then through responding to their adroit questions, challenges, and criticisms that I realized these disparate pieces needed each other for support, that they formed a whole. For these requests, comments, criticisms, provocations, and inspirations, I am extremely grateful. Particular thanks in this regard to Adam Rosenthal, Deborah Goldgaber, Armando Mastrogiovanni, Dorothea Olkowski, Niki Young, Michael Butler, and Ian Werkheiser, as well as the participants of the Limits of the Living Workshop, Engaging the Contemporary, and Digital Worlds Conferences.

This book would not exist without many, many conversations. For their generosity with their time and expertise, I thank Patrick Gamez (doubly so for the role his work played in the formation of this project and his comments on a draft of the manuscript), Anthony Kelley, Hannah Bacon, Jon Cogburn, Emily Cogburn, Arjen Kleinherenbrink, Edward Lenzo, Jonathan Wurtz, Rebecca Lenzo, Joseph Lucier, Jolene Zubrow, Joshua Kiesling, Ashley Kiesling, Joshua Dix, Joseph Mancinelli, Dana Flaharty, Kyle Keller, T.J. Tolton, Tom Sparrow, Rockwell Clancy, Kevin Taylor, the members of LSU's Ethics of AI Working Group, fellow participants of the Humanities Perspectives on AI Summer Institute (and to Rebekah Cummings and Elizabeth Callaway for organizing such an excellent program), the students in my What is Technology course, and many more I am no doubt forgetting.

I would also like to thank an anonymous reviewer, Christopher Coughlin, and the editors of the Sophia Studies in Cross-cultural Philosophy of Traditions and Cultures series at Springer for their comments, suggestions, and care with my work.

And as always, eternal thanks to my family for their limitless encouragement and support.

Contents

Chapter 1
Introduction: Power, Ethics, and Technology

> *"No book against anything ever had any importance; all that counts are books for something, and that know how to produce it."*
>
> *—Deleuze, "How Do We Recognize Structuralism".*

As thinkers, we live in the wake of the continental tradition's unwillingness to offer positive positions in both ethics and politics, especially in regards to technology. There are of course exceptions and caveats,[1] but as a rule, anything more detailed than a call for openness, generic resistance, or to await the event is seen as either naïve or hubristic. Both the left and right wings of the tradition embrace this, from Adorno and Horkheimer's negative dialectics which sees the activity of the thinker as entirely critical, to Heidegger's claim that only a god can save us and for this we must await the next revealing of Being. In the post-structuralists, it mutates into a near total ethical commitment. Foucault in particular makes this explicit in his view of the role of the intellectual, his aim to write in such a way not to prescribe any action but to make the present feel intolerable, or when he suggests that the response when asked what to do instead should be "it is not up to us to tell you the sauce with which we want to be eaten," (Foucault 2018, 194). In the fields which carry on the work of these thinkers, their retreat from norms became a norm of retreat. This left a vacuum in public intellectual conversation, especially in the United States, which

[1] Deleuze and Guattari, who this work builds upon, are examples of major figures that do not fall into this, as are Badiou, Fanon, Haraway, and Latour. In the realm of politics, this hegemony of critique is opposed by several continental strands of feminist philosophy, queer theory, and philosophy of race, and work that attempts to move beyond critique such as Felski, but the tendency is for the more positive aspects of these projects to be confined to specialists in subfields, being underappreciated and under-supported by both the mainstream of continental thought and the academy more widely (and often in those subfields themselves). In metaphysics and epistemology as well, new materialism, speculative realism, and similar movements oppose the hegemony of critique while being continually marginalized for this reason. Rarely, however, is ethics approached beyond the limits of critique in the continental tradition.

M. J. Ardoline, *Interiority Against Control*, Sophia Studies in Cross-cultural Philosophy of Traditions and Cultures 47,
https://doi.org/10.1007/978-3-032-16245-8_1

is currently being filled by the political fever dreams of a few tech bloggers and venture capitalists. Worse yet, many of these "thought leaders" are themselves making use of these very theorists for their constructive project. A picture of Foucault hangs on the wall at Peter Thiel's Palantir and venture capitalists now cite Deleuze and Guattari via Nick Land. While I appreciate the humility in refraining from prescription, I find the present moment intolerable.

This book asks how a good life is possible in the dawning of a new age of power, the age of control. My humble hope, then, is to find a few places in continental thought for the grounding of positive projects and new allies to help clarify these grounds. Such a project requires two stages: critical and constructive. These stages roughly correspond to the first and second halves of this book, respectively. The critical resources will for the most part come from the tradition of continental philosophy, especially the work of Michel Foucault, Emannuel Levinas, Bernard Stiegler, and most centrally, Gilles Deleuze. Here, the specific dangers of control will be diagnosed. The most pertinent of these dangers are, I will argue, the radical diminishment or even complete loss of self- and culture-shaping capacities. The constructive half will draw from the Confucian and Mohist traditions to synthesize them with the ethical and political aspects of Deleuze's thought. Both halves will draw on what has coalesced in the last few decades as the tradition of the philosophy of technology, such as the work of Albert Borgmann, Andrew Feenberg, and Shannon Vallor.

In this regard, I am engaging in a project called for by James Garrison in his *Reconsidering the Life of Power*, where he argues that continental thought has been successful in its self-appointed critical task, but has failed to provide satisfying constructive projects. The intellectual tradition most capable of rectifying this failure, his argument continues, is Confucianism. This is because both traditions share important, fundamental views on the relationship between human subjectivity and its socio-political context, but the Confucians develop a primarily constructive rather than critical project that follows from these views. Each tradition has its conceptual strengths and weaknesses, and so creative syntheses of them are required. I aim for a synthesis that focuses on Deleuze and Xunzi (with sundry other figures marshalled) to respond to problems at the nexus of contemporary technology, politics, and society, a nexus called control.

1.1 Power

Control is an undertheorized though popular term used by Gilles Deleuze. It names the new form of power made possible by information technologies alongside various political programs, techniques of shaping behavior, and shifts in economics. Deleuze's coinage is inspired by William S. Burroughs, but it also invokes cybernetics as the science of communication and control. This is apt because cybernetics forms a foundational vertebra in the backbone of information technologies, the techniques of power they make possible, and the self-understanding of power's workings that these techniques employ. It names a manifold strategy and growing list of techniques for influencing behavior that work via continual modulation of access, options, outcomes, and information. Control does not name a particular group, political or economic system, or ideology. Insofar as they aim to affect the conduct of others through tactics of modulation via information technologies, capitalists, anarchists, white supremacists, communist organizers, anti-racist activists, centrists, small business owners, doctors, our friends on social media, and so on all make use of control to some degree. Increasingly, we employ the tactics of control on ourselves as we explore new ways of self-shaping made available by information technologies such as fitness trackers. The forces of surveillance capital have embraced control most deeply and carried its development the furthest so far, but forms of power are ubiquitous in human interaction and control has not finished growing. As I write this book, tech oligarchs are in the process of gutting and redesigning the United States government in their image. If this succeeds, it will certainly mark a further mutation of control beyond the strictures of surveillance capitalism.

First, I should clarify that I am concerned with power in the Foucauldian sense. This term causes endless misunderstandings when Foucault is placed in conversation with political science and political philosophy more generally. In these domains, power is often understood from a bird's eye view as a top-down phenomenon where there are those on top with power and those below without it. Foucault will call this more traditional account a 'macrophysics' of power. He claims instead to be offering a "microphysics," paying attention to the minute flows of power (though like any good microphysics, the theory must be able to recover the macrophysical picture as we zoom out). As Foucault puts it:

> what defines a relationship of power is that it is a mode of action which does not act directly and immediately on others. Instead, it acts upon their actions: an action upon an action, on existing actions or on those which may arise in the present or the future. A relationship of violence acts upon a body or upon things; it forces, it bends, it breaks on the wheel, it destroys, or it closes the door on all possibilities, (Foucault 1982, 789).

Equally important is that the person acted on is "maintained to the very end as a person who acts," that is, they are not killed or otherwise rendered unable to take future actions, (Foucault 1982, 789). Power is conduct that acts on other's conduct. It is then a modally complex notion; it affects what one does indirectly by directly affecting what one might do. Violence and force are not power in this sense because they do not maintain the variety of possible actions.[2] Being bound and dragged somewhere against your will is not power, though one can struggle or not.

[2] This book does not deal with violence for the most part, not because violence is not the base of or constantly lurking behind power, but because as its base, violence is only the indirect cause or background condition of the effects and techniques of power I am interested in here. That some force of the state will shoot you if you do not comply is true of sovereign, biopolitical, and control societies. It is important, but it does not distinguish them. Furthermore, in the constructive chapters, violence plays little role in the good life, though power is inescapable. (those who wish to never exert nor have power exerted on them should move into a cave alone). Graeber 2016 critiques

Perhaps one wants to say that violence, force, or coercion is a limit point of power. To call the gradient from power to coercive violence a spectrum is not inaccurate, but it would be misleading as it implies 'shades of gray' which in fact do not exist even if we are not always capable of separating the two cleanly. Rather, this is a formal distinction in that violence is always lurking behind or within power (Foucault 1982, 789), but one or the other dominates a situation. This seems to give rise to a sorites paradox: when encountering a cop, at what moment between the cop hailing you and them pointing a loaded gun at you does the situation shift from power to force? Unlike the classical sorites paradox, this question has an answer, even if in practice we cannot locate the exact moment of the change. The reason for this difference is that the classical sorites paradox is additive, the change from grains of sand into a heap is merely the result of adding more sand. Moving from power to force is instead a phase transition, a radical change in the phase space diagram of the situation where possibilities collapse. Like water becoming ice, power crystalizes into violence. While there may be changes in degrees that lead up to unholstering, the act of drawing the gun is a freezing or boiling point, a singularity that becomes more than a mere difference of degree. I admit I struggled over the direction of this analogy. In the direction of water to ice, we see a sudden, drastic diminishing of degrees of freedom, much as coercion collapses the options available to the coerced. But for the direction from ice to water, violence or coercion risks a greater chaos, an entropic eroding of the complex structures of power. All else being equal, power is in some sense more stable and safer than violence.

The locating the inflection point between power and coercive violence requires attention to structure. Social position may change the nature of the interaction (e.g., given the history of racialized police violence in the U.S., the phase transition from power to coercion may be well before the cop draws their gun for black Americans, but at the moment of drawing for white Americans). Maintaining the structural situation but substituting people with different character traits may change what is likely to happen, how fast the boiling point is reached, etc., but it is not usually thought to change the nature of the interaction in the way social position does. However, we may wonder if a particularly stoic person finds themselves still in a situation of power under a credible threat to their life, or if a coward is nearly always coerced? Similarly, the existentialist or the pedant may respond that even in the case of a gun to one's head, one has the choice of compliance or sacrifice, and so because this does not determine one's conduct but only acts on it to make compliance much more likely, isn't this still power? If so, they say, then even though they do not survive as a subject capable of future action, wasn't that still their choice? If we make this distinction between power and coercive violence, is there really such a thing as coercion in this sense?

---

academics for ignoring violence and sees a sort of mystification at work in focusing on power or more metaphorical uses of violence. I agree with this critique and try to rectify this somewhat in Chap. 7 with a discussion of resistance and counterviolence. See Protevi 2025 for a Deleuzian, biocultural account of violence.

In lieu of an answer here, I will provide a definition to shore up as well as possible what is admittedly a heuristic distinction. When using "power"[3] then, I mean situations in which 1: there is a possibility space of what an actor can do, 2: some of the weights of the different possibilities (that is, how likely one is to choose an option) are downstream of human conduct and artifacts (the conduct need not be intentional), 3: that possibility space is more complex than a choice between being immediately harmed or not (otherwise this is force or coercive violence), and 4: the actor is likely to survive to make meaningful choices in the future. A rock in free-fall does not satisfy 1. A human being in free-fall might satisfy 1 insofar as they could either try to slow themselves through wind resistance by spreading out their arms and legs or getting it over with by assuming a diving position, but this would not satisfy 2. A hesitant sky-diver who is pushed would satisfy 1 and 2, but not 3, regardless of whether they manage to open their parachute or freeze (the latter case would not satisfy 4). Seeing the look of comfort (or disdain) on the face of the experienced skydiving instructor as one hesitates at the moment of their first solo jump satisfies all four. This would be true even in the case where something goes lethally wrong on the jump as someone making their first solo jump should be well-trained to the point that they were likely to have survived without issue. A lethal accident would not retroactively turn an expression of power into one of coercive violence.

A common critique of Foucault and his analyses of power is the lack of attention to more traditional accounts of power in terms of economics, class, etc. This criticism is fair, and rather than see Foucault as providing a total theory of political power, I have found it more useful to see his work as bracketing economic power and coercion (often even when he does discuss economics) to understand what is missed when we focus on the most obvious and most dominating forces in our society. That said, it is relatively easy to reconnect economic concerns to a Foucauldian conception of power by recognizing that a function of money is to command labor. Money, as the adage goes, is power, and this is precisely because money is a concretization of conduct that acts on others' conduct. This is one aspect of its reification of a social relation into an artifact. My goal here is not to give a total political theory that fuses Marx, Foucault, Deleuze, etc., but there are uniquely monetary techniques that are expressions of control which cannot be understood by bracketing economic explanations (for example, algorithmic wage theft as a form of power shaping subjects in addition to its use to increase profits via exploitation of workers).

Another such critique of Foucault's ideas and the way they've been picked up by scholars is they tend to imply a sort of idealism, a discourse monism. I think it unlikely that this was Foucault's position or that it follows from his central ideas given the amount of time he spends analyzing the artifacts that shape and make possible expressions of power. He does not, however, give a metaphysical underpinning to his account, and this coupled with how his work is often taken up has led to this idealist misconception. Connecting this account to Deleuze's metaphysics will

[3] This usage also has nothing to do with power in the sense of a powers ontology that I defend in Ardoline 2024. When discussing powers in that sense, I instead use here capacities or abilities. This difference roughly tracks that in the French between *poivre* and *puissance*.

provide a materialist grounding for it. Here we should also be concerned with a rhetorical issue that causes a further misunderstanding. When one focuses on the critique of power and avoids a constructive project, there is the implicit implication that the best course of action is to release power, to minimize it, to undermine, to not be a node across which power flows, and to above all avoid exerting power ourselves. We may call this error the "beautiful soul" interpretation of Foucault. It equates power with danger and harm alone (despite Foucault's claim that power can be productive just as much as repressive), and it implicitly recognizes that power flows in all human interaction. How could two people be a part of each other's lives without acting on the conduct of each other? One cannot be a good friend if one does not allow power to flow through the friendship in some way. A duplicitous Rousseau-ian demon possesses an otherwise useful tool. Instead, our approach will be to understand how power functions in the process of shaping subjects and groups so we can better intervene in these processes of which we are a part. In the constructive portion of this book, we then see how power is enabling and where control provides new possibilities for our capacities and joys alongside its dangers.

That said, our concern here is not power in general, but with the rise of a new age of power. There are always uncountable ways in which power flows through human interaction, but, so Foucault's program contends, one can discover certain organizing principles or strategies that make up a dominant form of power for a time and place. These are the well-known categories of sovereign, disciplinary, and biopower, among other contenders. Here, Koopman cautions us not to think of power in terms of epochs, (Koopman 2019, 171–2). Instead, he describes layers of power, where new techniques and technologies sediment into strata atop previous layers, sometimes upsetting or breaking apart that which is underneath it. This is a useful image as it is helpful for resisting the image of periods of power as discrete temporal moments. These periods do not mark a clean break (sovereign power's techniques, strategies, and justifications persist into the age of biopower), instead, "resisting notions of replacement and obliteration, Foucault emphasized interpenetration and superimposition," (Erlenbusch-Anderson 2018, 31). Nor do they describe every expression of power that happens within that time or place (there are always actions on the actions of others that do not proceed according to the logic of either the current dominant or preceding strategies of power; no dominant strategy is total). With these concerns in mind, I will still talk of ages or epochs of power for the same reason that paleontologists mark eras through radical shifts in what appears in the strata they explore. This also makes obvious why this book will not use Foucault's methodology despite using his concepts. Whereas Foucault undertook histories of the present, so unearthing the strata on which we stand, this work is focused on the surface, the topsoil as it becomes sedimented.

1.2 Control

A central claim this book defends is that we have entered a new age of power, which Deleuze named "control." Like anything, control's essence is first hidden, mimicking other things until it is strong enough to show itself. This is doubly so for a strategy of power, which Foucault reminds us is tolerable only insofar as it can partially conceal or disguise its operation, (Foucault 1990, 86). That we have entered a new epoch of power means that it has at least begun to show itself to a degree we can name and analyze; it does not mean that control is at the height of its abilities or has even finished cohering.

Control names a form of power that operates by the continual modulation of access to resources, spaces, information, activities, and so on, (Deleuze 1997, 179–181). We can imagine this as a mansion in which each room has several doors leading to other rooms. By systematically alternating which doors are locked and unlocked, we keep a wanderer traveling throughout the mansion in such a way that they are always able to move and always have options. However, there may be huge parts of the house which remain entirely unreachable through locked doors, hiding whatever is behind them by preventing the thought of entering them from even appearing as an interesting option. The constant availability of options of unlocked doors promotes a feeling of freedom. The wanderer continually moves, opening new doors, maybe trying some locked ones and then moving on. Some of these locked doors become unlocked and some unlocked doors get locked, prolonging the sense of discovery and novelty. However, this makes the doors which remain locked, which never open, nearly invisible. There are always other options. The locked doors are never seen as an impasse to overcome; there is no reason to try to break down these doors to see what waits behind them because there is always somewhere else new to go. This amounts to the continual shaping of behavior by the modulation of novelty, movement, and desire, minimal violence required (which is certainly not to say the actual violence used is ever minimal).

I use the term "control" for 3 reasons. First, I think Deleuze's description is prescient and useful, despite being thoroughly incomplete. In fact, that the account is incomplete is useful as control had not shown itself fully in the early 1990's when Deleuze named it (and I think it still hasn't). Other attempts to understand our age as one of data-power, psychopower,[4] etc., are too determinate and too narrow in their focus. Second, control is general in that it attempts to name not a particular crisis, but what is behind the crises we face. For example, something not yet fully understood, something that might not yet be knowable is happening in political economy. We see this in that the attempt to categorize this change has provoked accounts as far apart as claiming it is a new, rogue stage of capitalism (Zuboff 2019) to claiming it has overthrown capitalism and instituted a new form of feudalism

[4] See Camp 2012.

(Varoufakis 2024).[5] Meanwhile, it gives rise to new ideologies of cybernetic monarchism (as seen in the so-called 'political philosophy' of Neo-Reaction) and a renewed hope for Marxism and anarchism alongside Left Accelerationism or Prometheanism (an admirable attempt at a constructive politics in the continental tradition, currently on life support). What these accounts all share is they are a response to control, whether to understand it, accelerate it, subjugate it for (post-) humanist ends, or flee from it into a cowardly and violent reactionary fantasy. This is not to say that a single unified source is driving all current interesting political change or experimentation (for example, I am doubtful that Rojava is in any deep sense a response to or a manifestation of control, even if its effects are felt there too), nor should we think of control qua cohering strategy as a single "source" in its own right.

There is, I admit, a fourth reason why I use "control." No shortage of neologisms have been coined in attempts to theorize the new age of power that we all sense crystalizing around us. We are told we have witnessed the rise of data-power, cyberpower, psychopower, neuropower, and so on. If all these terms do in fact name new dominant strategies of power, then we must posit an unbelievable increase in ways of influencing and commanding. How long was the epoch of sovereign power? How thick its strata? Have we really accelerated from the centuries of biopower to a new epoch taking hold every other week which are yet somehow all related to information technologies? Are we truly experiencing an "explosion of powers," (Gamez 2023, 90)? More likely is that rather than picking out a dozen different competitors, they all name components, tactics, precursors, and other bits and pieces of this newly sedimenting strata of power. The proliferation of terminology strikes me as equally an expression of academic marketing, with each scholar establishing their brand name. This branding exercise both recapitulates a central tactic of this new form of power (fragmentation via excessive proliferation and novelty), and it impedes the work of actually understanding control (or whatever one calls it) by carving it into opposing accounts. Rarely does this carving happen at the joints. I have benefitted greatly from the work of these scholars and see this not as a failing of theirs but as resulting from the realities of contemporary academia where every insight must be sold as revolutionary, every new analysis of power must be epochal no matter how specialized the subfield, and justifying one's work includes pleasing marketers. And so, I choose a term already in use, one with wide-enough scope to potentially be adequate to what we are attempting to name, connected to a thinker popular enough to satisfy the marketers and which is attached to an ongoing research program already resisting fragmentation, in the hope that it allows me the space to not pretend to philosophical greatness but to work as an intellectual tinkerer.

I do have some misgivings about the term that are best to say upfront. First, control seems to imply top-down power. We will see in Chap. 3 that control instead is as likely to be horizontal or bottom-up as it is top-down, if not more so. Second, the

---

[5]This work mainly makes use of Zuboff's framework, not because I take it to be a complete account of socio-economic situation or without its flaws, but because it more thoroughly engages with the relation between the current political economy of control and the formation of subjects.

singularity of the term makes control sound like a vast, unified network. This may be exacerbated because this book will discuss some particularly powerful nodes of control. However, power is a diffuse phenomenon, one that can operate without intention behind it. Power is conduct that acts on others' conduct. If I open a browser to google some new product I have just heard of but then decide not to because I would rather not be bothered by ads about it later, this is a minor, relatively uninteresting example of control without anyone "controlling" me. Third, it is a common enough word that perhaps a neologism would remove some ambiguity, though I have not found one that captures the phenomenon in its entirety. Finally, "control" has a menacing air to it, especially due to the connotations of its colloquial rather than technical use (Koopman 2019, 169 rightly notes the sense of paranoia around the term as well). However, this last concern is true in some degree of the various other terms for forms of power. As discussed above, this leads to a knee jerk reaction that power is bad. This is common not just among casual readers or critics of post-structuralism but some of its more important interpreters as well. The most prominent example in recent memory is Agamben's shameful response to Covid-19 mitigation measures under the aegis of opposing biopower. Just as there are techniques of biopower we would embrace or prefer to live according to rather than what preceded it, there will be grand opportunities made possible by control, and perhaps much to prefer about it despite its dangers.

1.3 The Good Life

On this latter note, this book aims to discover conditions for the good life, or rather, good lives, from within control. In doing so, it joins a recent, growing discontent with a dominant approach to politics in continental philosophy of the last few decades. This discontent is found mainly with approaches to political thought that diagnose the functioning and dangers of power impressively, but in response, counsel mere resistance or melancholy. This is doubly disappointing in that post-structuralists are the first to point out that resistance often does not interrupt power, but rather makes it flow. Here, the above-mentioned *Reconsidering the Life of* Power is particularly compelling. In it, Garrison argues that there is a way out of this: synthesize continental philosophy with the Confucian tradition. The main reason for this is that both traditions share a deep concern with politics as well as fundamental claims about the nature of subjects. Both traditions take subjects to be shaped by their social, political, and economic situations, as well as through aesthetics and technology. The difference, however, is that the continental tradition sees in these facts our passivity and capture in the face of the relations that form us whereas the Confucians see these conditions as that which makes our freedom, goodness, and the like possible in the first place. This is particularly relevant to any critique of control because control works by disrupting these conditions of subject formation. If so, then control contests us at the very site of agreement between these two traditions. The synthetic philosophical project that Garrison argues for is then uniquely qualified to address the harms this new strategy of power poses to the possibility of good lives.

There are at least five components of the good life that are under threat by control. These are interiority, activity, self- and social cultivation, and creative resistance. If there is a unifying ground to these five components, it is desire. Control operates by incredible predictive power that gradually offloads and disrupts the formation of subjectivity via organizing desire. Consciousness is a response to problem situations that cannot be well-solved by fast twitch responses (this claim is defended in detail in Chap. 2 through Simondon on psychic individuation). The more the prediction-stimulus-reward loop tightens around us, the less need for consciousness, for interiority itself. In doing so, we become ever more passive, ever more consumers rather than creators. Without activity, we lose the ability to shape ourselves, our social connections, and our cultures. Finally, our very resistance to this situation becomes predictable, manageable. Each of these components will be explored through a different thinker. Interiority through Levinas, activity through Deleuze, self-cultivation through Xunzi, social cultivation through Sylvia Wynter and Li Zehou, and creative resistance through the first great critic of the Confucians, Mozi. My aim is to synthesize these disparate projects into an account of how good lives are possible in the epoch of control. Admittedly, I think most of the resources to do so are already found within Deleuze's philosophy, but these resources are best drawn out in encounters with other thinkers.

Control is not synonymous with digital or information technologies. Modulating rates of debt is a paradigm technique of control that can be done with paper spreadsheets, though it is made easier with information technologies. Such technologies are not inherently inimical to the good life. Nor is there anything inherently positive or neutral about them. Artifacts are active structures of capacities consisting partially of concretized values and knowledge (human or otherwise) which can molded, mitigated, and radically transformed by their relationship to other things, just like humans. Because power is not merely top-down, there are liberatory uses of these technologies as well as of control. Similarly, the breakdown of the subject is required for any transformation, so the inherent threat to subjectivity by control is itself useful despite its danger. It is not, however, neutral. If the threat to the subject is total, it prevents transformation insofar as it prevents the formation of new subjectivities. This is as harmful to human flourishing as enforced stagnation.

This book is concerned with how to live well with that fact, not with how to destroy surveillance capitalism (which we should, especially the capitalist part, contra Zuboff), not how to live a life free of power (which is an incoherent fantasy), nor how to reverse the rise of control (an ambivalent dream at best). The dangers of control are not some *telos* of this epoch of power. History has no *telos*. Rather, these dangers are a worst-case scenario of a particular form of power coming to colonize the majority of human interaction. The dangers of control are not inevitabilities. Some of them I take to be physically impossible though they are limit-cases of an otherwise possible path we may travel. For instance, I am not sure that human interiority can ever be completely eradicated while we remain the same biological creatures we are (in the long term, we won't remain such creatures, but this is a book about the short term[6]). The dangers of control are instead a singularity in our species' phase space: it is what, given the conditions of human life, technological capacity, and political reality, we are pulled towards. It is not the only such singularity, and it is not the worst one, though it may be the nearest. The only way to remove this singularity is to remove its conditions, and while controversial, I would prefer that both the internet and humans and their environment, remain, change each other, and ultimately flourish together. More than ever, our politics and our ethics cannot be based on a faith in human exceptionalism.

Finally, this book has a commitment to some form of techno-optimism. This term has recently been claimed by right-wing accelerationists, Nick Land and venture capitalist Marc Andreessen in particular. But these projects are in no sense techno-optimist; they are capitalist-optimists, calling for the deregulation of markets, greater corporate power, and lacking any attention to how technology may contribute to flourishing, human or otherwise. Instead, what I take techno-optimism to be is the commitments that 1) it is through technology that we can explore, take an active role in, and make more durable the conditions in which humans flourish (along with non-humans) and 2) that human ethical and political improvement does not come from will, action, or discourse alone, but from the technological scaffoldings which we build around ourselves, the knowledge and cultural memories we retain in them, and the habits we take on through them. Aristotle's great lesson here is that technology and ethics, or rather *techne* and *arete*, are united in that they both aim to make our flourishing more robust against *tyche*, the whims of chance. *Techne* is an *arete*; *arete* is a *techne*.

Despite all the understandable reasons to be critical of the effects of industrial and information technologies on the world, one can't help but psychoanalyze the impulse towards techno-pessimism as one more manifestation of the repression of human's dependence on technology that Stiegler identifies (Stiegler 1998). In short, he argues that what makes humans human is that we transmute individual memory into cultural memory through concretizing it into artifacts. Writing is of course the most obvious example, but any artifact embodies some of the memory, some of the world of those who made it such that it can be passed on to others. A hammer concretizes an individual's knowledge, experience, and skill in the same way as, though to a lesser degree than, a journal. Were it not for these technological scaffoldings, improvements of the human condition could not be preserved.

The belief in history as inherently Progressive, from the Hegelian view of History through all its successors, obscures this fundamental fact. Positive changes will disappear if they are not maintained, further concretized, and made robust. But if so, then such positive changes are only as durable as the artifacts that embody, support, and reproduce them. This is a frightening thought. The technological concretization of progress (importantly lower case) is spiritualized by Progressives in a way that hides its true nature. To soothe one's fears of potentially losing what has been gained

[6] For an excellent analysis of what we may become, see Roden 2014; for an imaginative exploration of divergent paths of human evolution, most of which lose higher-order cognition, see Nemo Ramjet's *All Tomorrows*.

and what one comes to depend on (especially when that dependence is on the concretized labor of anonymous others), recognition of the material is repressed into something spiritual. Artifacts can be destroyed; History cannot. In attributing to History what is the role of technology, we obscure the fact that we must maintain these systems (Stiegler generalizes this critique to many other idealisms as well: Platonists fear that libraries may burn, so they soothe themselves with the fantasy of knowledge retained in the eternal Forms). This is a fact that reactionary forces, never being duped by the idea of the inherent Progress of history, know well and use to their advantage. Any leftist project that represses the necessity and vulnerability of technology in favor of a spiritualized historical progress leaves itself unprotected against these attacks.

Techno-optimism does not amount to the belief that technology is inherently good or is itself necessarily progressive or liberatory. Artifacts must be integrated well, but this very integration is a form of social technology: it is secondary instrumentalization or ritualization. What Heidegger attributes to the essence of technology, I follow Feenberg in attributing to the economic capture of rationality. The tinkerer, the mechanic, and the hacker know nothing of *Gestell*; one who shows proper care and respect for what they build and rebuild does not reduce their devices to mere means. In the last instance, our care for our artifacts and their social integration is care for and of our selves, others, and our worlds.

1.4 Chapter Summaries

We will begin with a problem that I take to show control in its idealized form: the exhaustion of subjectivity. This is the condition where the interiority of the human mind is made unnecessary through the ever-expanding ubiquity of information technologies and their powers of recording, stimulation, and prediction. Whether this danger is fully realizable or not (and we should not be quick to discount it), it shows control at its limit. In doing so, we see both a pertinent danger and control in its purified form. This latter insight allows us to better diagnose it. First, it recounts Gamez's description of the problem, and the work on surveillance capitalism that it is based on. Second, it argues that Simondon's account of psychic individuation provides a plausible mechanism for this exhaustion and thereby gives us an understanding of how it could take place. Third, it is argued that the exhaustion of subjectivity is an example of the wider phenomenon of technological deskilling, in particular of moral deskilling as described by Vallor. This then connects the exhaustion of subjectivity with work on subjectivation in continental political philosophy and the virtue ethics tradition.

Chapter 3 enters into the debate over whether the form of power exemplified by information technologies is an extension of biopolitics or if it is discontinuous with it. I argue in favor of the latter, and that this new epoch is best characterized by what Deleuze calls control. This argument is based on recounting the essential features of biopolitics that Foucault elucidates (the management of life, focus on individuals and populations, cultivation of discipline, and the use of statistics) and showing this contemporary form of power shares none of these (instead, it manages access, focuses on dividuals and masses, eliminates discipline, and is made possible by linear algebra). This then makes the case that the exhaustion of subjectivity is a unique danger of this new form of power.

Chapter 4 builds on Chaps. 2 and 3 by exploring what harms result from the loss of interiority that the exhaustion of subjectivity would amount to. Both Deleuze and Foucault (and post-structuralists in general) have well-known critiques of the subject, so why would we not celebrate its exhaustion? Here I argue that in his late work, Deleuze recognizes that what he took to be inherently liberatory, especially in his work with Guattari, is also what makes control possible. This means the critique of the subject must be reappraised. It is argued that the concerns over the problems of subjectivity in terms of autonomy and the atomic individual are well placed, but that subjectivity-*qua*-interiority need not be thought of in these terms. Here I argue that we can maintain that critique while defending several important aspects of interiority, in particular the Kantian view that interiority is a necessary condition of society and the Levinasian view that it is also the condition of truth, desire, and ethics. This then marks the shift towards the positive project.

Chapter 5 argues that it is possible to become active in one's own subjectivation. That is, one can cultivate their interiority. This is defended first through a synthesis of Deleuze's account of individuation (seen in Chap. 2) and Xunzi's account of self-shaping through artifice. This is a virtue ethics, but one that avoids Aristotle's equation of the good with self-sufficiency (which is critiqued in this chapter). This culminates in an account of what it means to construct a *dao* or path for oneself. By understanding the role that social institutions (understood as Confucian rituals), art, and (especially information) technology play in shaping one's self, it becomes possible to order relationships between these things such that they reinforce one's aims of self-cultivation.

As shown in previous chapters, it is not enough to have an individual solution because control shows us is that we have never been atomic individuals. This chapter addresses this concern by turning to the cultivation of social groups, or what Deleuze and Guattari call in *What is Philosophy?* the summoning of a new people. Insofar as a *dao* is already social, some account of how the social itself can be reshaped by the formation of rituals rather than merely reproducing the subjects of power is needed. This chapter explores this issue by looking at two other accounts of the formation of peoples, Sylvia Wynter's sociogeny and Li Zehou's subjectality, and the role technology plays in both. Ultimately, it is argued that becoming active in our social individuation happens through the materially engaged cultivation of beauty.

The positive project above raises the question of what forms of resistance are justified to the current ways information technologies, and their socio-economic deployments, are shaping ourselves and our groups. I first give Simondon's account of technical individuation to explore how humans can intervene in shaping technical objects. This is expanded through Feenberg's schema of secondary instrumentalization. These accounts together provide a way of separating resistance that merely stabilizes power from resistance which disrupts it. I use this distinction to respond to Zuboff who argues that resistance is capitulation to surveillance capitalism (and so to control). To provide an ethical principle that clarifies this, I turn to the ancient Mohists, their principle of benefiting the world, and their opposition to offensive warfare through counter-siege engineering. It is then argued that technological pessimists and quietists are mistaken; control may only be made ascendant through information technologies, but resistance to control does not mean resistance to such technologies in general. Instead, it requires technological innovation in its own right that changes how power can be expressed through those technologies.

The Conclusion summarizes the systematic account given. It then asks what a good life may look like assuming that this new form of power is here to stay. It is argued that while control challenges the conditions of subjectivity itself, it is also this capacity that opens up the possibility of radically reshaping ourselves and our social connections.

1.5 A Note on Systematicity

This project makes use of a long list of proper names; it attempts to piece together insights of the great thinkers of several traditions. This is not uncommon, but it is always a risk. Insights separated from their context may appear to sit well together until one traces their conditions and implications only to find a tangle mess of contradictions. Philosophy does not merely collect or create true or compelling ideas; it must always attempt to synthesize these ideas so that they may be saved from their context and hang together. That is, it must systematize. This synthesis is explicit where I bring together continental and ancient Chinese traditions, but it is implicit in bringing together thinkers within the continental tradition. There is a clear zone of overlap in the projects and influences of Deleuze, Simondon, Foucault, Stiegler, Wynter, and so on (Levinas is something of an exception here, which I address in Chap. 4), but there are also fundamental disagreements. In explaining an insight, I make use of the thinker who I think has given the best or most direct account relevant to the task.

Systematically, I remain in this project committed to a metaphysics of difference à la Deleuze whose foundations I partially worked out in Ardoline 2024. This means when making use of the insights of other thinkers, I am obligated to maintain a set of constraints, and to use those insights only in ways consistent within them. Metaphysically, these include the priority of difference, a non-reductive form of materialism, and immanence. More importantly for this project, Deleuze's metaphysics comes with an implicit philosophy of technology which has been admirably unearthed by Smith 2024. While I will explore these underlying commitments to a particular metaphysics of technology throughout, I list the following as central commitments of this project, all found in Smith's account. First, technology is prosthetic in that it externalizes, extends, and scaffolds human capacities. Second, biology is already proto-technical; that is, biological organs are already a sort of proto-prothesis. In other words, the difference between artifact and natural organ is less a metaphysical difference and more a difference in construction, material, location, etc. Third, what characterizes human created artifacts, that is, technology as we commonly use the word, is that they enter complex enough relations with humans, their environment, and other artifacts such that they have their own independent paths of evolution, importantly different from natural selection. Technical individuation is then a sort of "exodarwinism," a term Smith draws from Serres (Smith 2024, 377). Fourth, our understanding of the world is more fundamentally a form of maker's knowledge rather than propositional knowledge. Maker's knowledge involves both the idea that we know what we can make, but also that we learn from what we make and use. Knowledge can be concretized in and generated by artifacts.

Where a potential inconsistency may be a problem, I attempt to adjudicate it and make clear the synthesis. Where I think an insight is available to or within this metaphysics, but is said or explicated better elsewhere, I make use of it without comment.

References

Ardoline, Michael J. 2024. *Deleuze, Mathematics, Metaphysics: Difference and Necessity*. Edinburgh University Press.

Deleuze, Gilles. 1997. *Negotiations 1972–1990*. 1st ed. Translated by Martin Joughin. Columbia University Press.

Erlenbusch-Anderson, Verena. 2018. *Genealogies of Terrorism: Revolution, State Violence, Empire*. Columbia University Press.

Foucault, Michel. 1982. The Subject and Power. *Critical Inquiry* 8 (4): 777–795. https://doi.org/10.1086/448181.

Foucault, Michel. 1990. *The History of Sexuality, Vol. 1: An Introduction*. Vintage.

Foucault, Michel. 2018. The Analytic Philosophy of Politics. Translated by Giovanni Mascaretti. *Foucault Studies*, June 29, 188–200. https://doi.org/10.22439/fs.v0i24.5532.

Gamez, Patrick. 2023. Inhuman Hermeneutics of the Self: Biopolitics in the Age of Big Data. *Foucault Studies* 34:80–110. https://doi.org/10.22439/fs.i34.6939.

Graeber, David. 2016. *The Utopia of Rules: On Technology, Stupidity, and the Secret Joys of Bureaucracy*. Melville House.

Koopman, Colin. 2019. *How We Became Our Data: A Genealogy of the Informational Person*. 1st ed. University of Chicago Press.

Protevi, John. 2025. *Regimes of Violence: Toward a Political Anthropology*. Univ Of Minnesota Press.

Roden, David. 2014. *Posthuman Life: Philosophy at the Edge of the Human*. 1st ed. Routledge.

Smith, Daniel W. 2024. Deleuze and Technology. *Deleuze and Guattari Studies* 18 (3): 373–392. https://doi.org/10.3366/dlgs.2024.0563.

Stiegler, Bernard. 1998. *Technics and Time, 1: The Fault of Epimetheus*. 1st ed. Stanford University Press.

Van Camp, Nathan. 2012. From Biopower to Psychopower: Bernard Stiegler's Pharmacology of Mnemotechnologies. *CTheory*, Accessed May 9, 2012.

Varoufakis, Yanis. 2024. *Technofeudalism: What Killed Capitalism*. Melville House.

Zuboff, Shoshana. 2019. *The Age of Surveillance Capitalism: The Fight for a Human Future at the New Frontier of Power*. Public Affairs.

Chapter 2
The Exhaustion of Subjectivity

All things emerge at midnight, shrouded in darkness and lacking the power to show themselves as something new. They may be anticipated before this, but they can only be ktnown once they appear in the light of day. The dawning of control means we can begin to glimpse its potential dangers even if only partially. The most pertinent of these dangers we can currently glimpse is what Gamez calls the exhaustion of subjectivity (Gamez 2023). This is the condition where the interiority of the human mind is made unnecessary through the ever-expanding ubiquity of information technologies and their powers of recording, stimulation, and prediction. This chapter explores the nature of this problem. First, I recount Gamez's description of the problem and the work on surveillance capitalism that it is based on. I then use Simondon's account of psychic individuation to explore a possible mechanism of this exhaustion. Third, I argue that the exhaustion of subjectivity is an example of the wider phenomenon of technological deskilling. This both grounds its possibility and allows us to connect our concerns with work on subjectivation in continental political philosophy and virtue ethics.

2.1 Surveillance Capitalism and the Exhaustion of Subjectivity

The most recent steps of control into the light of day have been via digital platforms and massive tech corporations. Shoshana Zuboff calls this surveillance capitalism. While there are many analyses of the contemporary situation,[1] Zuboff's remains the most useful for understanding the exhaustion of subjectivity (Gamez 2023). Whether

[1] For a radically opposed but viable account, see Varoufakis 2024 which claims that capitalism has been overcome and we have entered techno-feudalism. This is a claim that I think is currently not

M. J. Ardoline, *Interiority Against Control*, Sophia Studies in Cross-cultural Philosophy of Traditions and Cultures 47,
https://doi.org/10.1007/978-3-032-16245-8_2

surveillance capitalism correctly names the age we are in or not, the techniques of power and the aims that Zuboff analyzes are at least a major portion of control at its current stage of development.

Zuboff offers several definitions of surveillance capitalism, the most direct of which are that it is "a new economic order that claims human experience as free raw material for hidden commercial practices of extraction, prediction, and sales," "a parasitic economic logic in which the production of goods and services is subordinated to a new global architecture of behavioral modification," and "a movement that aims to impose a new collective order based on total certainty," (Zuboff 2019, v). This economic structure operates through the collection and commodification of enormous stores of data in order to predict and control behavior (often but not always purchasing behavior) so that it generates ever increasing profits. As control becomes a more general strategy used by non-corporate actors, it will come to serve other ends in addition to profit. The means, however, will maintain the same general diagram.

Every action we undertake that is mediated through a computer (in the broadest sense) leaves traces. These traces can be formed into data that encodes our behavior. Such data is ripe for further use in training machine learning, building predictive user profiles, and so on. Zuboff calls this data "behavioral surplus." Surveillance capitalism begins with the realization that behavioral surplus is itself a useful and lucrative commodity. More and more minute details of the user's life are stored, bought, sold, and analyzed to make that user's behavior legible, predictable, and influenceable. In terms we will explicate below, this means enormous details about individual users' lives are entered into technical memory, and so accessible and usable by others. Often, data is collected and stored without any use cases in mind under the belief that it will later become useful. The use of behavioral surplus and machine learning ultimately results in autonomous algorithms that aim at behavioral modification.

Zuboff sees this as an aberrant mutation in capitalism that, with some effort, could be reversed. In doing so, her goal in opposing this mutation is a return to the capitalism of the mid-twentieth century. I do not see how this is possible nor why one would want it to be, especially given that the techniques of surveillance capitalism extend a project of deskilling as old as capitalism itself (which is explored below). Furthermore, while the main use of these technologies of prediction and modification has so far been in advertising and modulating the use of platforms, events like the Cambridge Analytica scandal of 2016 and government contracts with corporations such as Palantir hint at much wider uses. Merely reversing course on capitalism will not stymie control; nothing about, say, a theocratic technofeudalism is incompatible with control.

In "Inhuman Hermeneutics of the Self," Patrick Gamez analyzes what the capacities of such technology organized under this new economic imperative

correct, but depending on how the tech-oligarchic takeover of various governments goes, will retroactively have been true.

means for Foucault's account of biopolitics and for the self. Here Gamez has two central theses. First, that even in the dawning age of surveillance capitalism, we are still within the regime of what Foucault calls biopolitics. Second, that surveillance capitalism makes possible what he calls the exhaustion of subjectivity. The first thesis amounts to the claim that surveillance capitalism ultimately works on the two biopolitical principles of statistics and the management of populations. We will return to this claim in the next chapter, whose goal is to argue that biopolitics does not account for this mutation. The second, the exhaustion of subjectivity thesis, is that if our actions are completely observable, predictable, and malleable by the forces of surveillance capitalism, then human action becomes mere behavior. Behavior is human action emptied out of "unnecessary" posits like desire, belief, and ultimately, interiority itself. The logic of such a program is that "there is no longer a need for interiority or to see our actions as the expression of an inner truth; the correlation of traces is all that is needed to predict our behavior to a frightening degree," (Gamez 2023, 106). Here I agree, and see this as one of, if not the, most important potential dangers for diagnosing our current condition.

While the exhaustion of subjectivity is a recent danger, concern with technology's potential to damage subjectivity goes back at least to Heidegger, who in "The End of Philosophy and the Task of Thinking" claims that cybernetics is the end of philosophy (Heidegger 2008). It is the end both because it is the culmination of the tendency of Western philosophy to give rise to specialized sciences, and because with the success of these specialized sciences, the problems of philosophy are not so much solved as made irrelevant. Because cybernetics is supposedly domain general (i.e., it can be applied to any objects or systems), Heidegger argues it is the final science needed to allow the sciences to communicate among themselves without recourse to philosophy. This follows from his claim that the essence of technology is enframing (*Gestell*), or the deep structure of our sense-making that leads us to see objects only as their instrumental use, or ultimately, as mere stores of energy to be made available for our projects (i.e., as standing reserve), (Heidegger 2013).

As the general science connecting all the others, cybernetics gives a way of reducing the problems raised by trying to make sense of different sciences to problems of efficiency in communication and control. The sciences, and so human thinking itself, become one more thing understood only as available energy to direct towards purely means-end reasoning. The sites of conflict between the sciences that might then provoke deep philosophical thinking are covered over in advance. Cybernetics completes enframing. Such a situation would be the triumph of instrumental thinking in that one no longer encounters the sort of complex breakdowns in our everyday going about the world that provoke philosophical reflection.

There is something to Heidegger's concern. For example, a similar risk was recognized by the father of cybernetics himself, Norbert Weiner, prompting him to write *The Human Use of Human Beings*. Weiner is however, more optimistic than Heidegger that good design can avert these problems.[2] I won't recount here how the information age comes about through the constellation of cybernetics, Shannon information, the work of Turing, Hayek and neoliberal economics, and management sciences as this would require its own study, several of which now exist.[3] However, the arc of this history of extraction from Charles Babbage's work in both computation and the division of labor through management to contemporary artificial intelligence is summed up adequately by Pasquinelli 2023's declaration that "labor is the first algorithm," (Pasquinelli 2023, 16).

Heidegger's account has several flaws when used to diagnose the damage cybernetics does to subjectivity, of which we highlight three. First, the historical sweep and lack of specificity of his transcendental account gives us little with which to get a grip on the actual processes by which the end of thought is carried out. If this is a process set in motion by the ancient Greek understanding of Being, then why does it accelerate only when tied to the profit motive? This is not a question he can answer, even if he occasionally gestures at economic forces. The supposed ontological force of enframing makes it explanatorily prior to such forces. This also obscures attention to particular examples of technologies by treating them as all a part of one narrowly directed historical force called technology.[4] Second, it is premised on a particularly elitist view of thought wherein only philosophy truly thinks; the sciences do not or do so only in a derivative sense. Presumably, most people do little of what Heidegger would consider philosophy. Recall Jaspers's anecdote that Heidegger thought Germany would need only two or three professors of philosophy. When Jaspers asked who he had in mind, Heidegger remained silent. Whatever Heidegger means by thought here, it is certainly a much more rarefied activity than the general interiority we are concerned with. Third, as a combination of the above, the broad focus on this enormous world-historical sweep gives us no explanation of how this operates on the individual level.

---

[2] One wonders whether their judgments on technology and if these issues are surmountable are downstream of both thinkers' political concerns, with Heidegger's pessimism following from his fascism, cultural chauvinism, and worry that increased communication is related to a leveling down of potential greatness (who can hear his pining for a god to save us and not wonder if "god" is a euphemism for Führer), whereas Weiner sees a positive continuity between democracy and cybernetics through increased social feedback, and for example, chastises the U.S. for failing to realize this ideal of social feedback so long as white supremacy remains (Weiner 1971, 50).

[3] In addition to Zuboff 2019, see for example, Mirowski 2006, Pasquinelli 2023, and Gamez 2025.

[4] See Verbeek 2005, Ch. 2. Verbeek's critique does not go far enough in this regard however. Verbeek's own account follows Heidegger's reduction of artifacts to our possibility of encountering them. Kleinherenbrink 2024 argues this makes object-object relations independent of human experience of them unthinkable. Thinking such object-object relations becomes ever more pertinent as more of our world becomes automated.

More useful to us in these regards is Arendt's claim that "The trouble with modern theories of behaviorism is not that they are wrong but that they could become true," (Arendt 1998, 322). That is, human action will be mere behavior that is explainable (and predictable) via external stimulus alone. Zuboff herself echoes this concern noting the role of behaviorism in the development of surveillance capitalism's techniques of prediction and intervention, but the process and its relation to cybernetics is already laid out in Galison 1994. The behaviorist program is mutated by cybernetics. "While prewar behaviorists might have cautioned against the ascription of internal states, war made it impossible; reading the hidden enemy meant reading his actions. In the mechanized battlefield, in those life-and-death confrontations with an enshrouded enemy, the identity of intention and self-correction was sustainable, reasonable, even 'obvious,'" (Galison 1994, 263). This new concern for the intentions and opacity of the enemy was not a sudden respect for the ineliminable depths of the human mind, but a rendering of those depths into a problem of visibility to be solved. While prewar behaviorism aimed to explain human action without recourse to anything interior, its cybernetic variant aimed to render the interior visible and predictable. What is completely transparent ceases to be interiority and becomes a legible surface. Behaviorism was no longer a research methodology; instead, it became an engineering goal.

In the post-war world, this intellectual project undergoes another series of symbioses as it enters the realm of economics. Most important here are the encounters with advertising (which, in a certain sense operationalizes psychoanalysis through Edward Bernays' reconciling depth psychology with its great enemy of behaviorism) and the economic project of neoliberalism. This latter encounter is perhaps the apotheosis of this tendency even if it must wait a few decades for it to incarnate itself in global information technology networks, wherein the market becomes understood as a cybernetic system of global coordination. Neoliberalism worships the market because the market delivers what gods only promised: the subjugation, modulation, and judgment of all human behavior. In fairness to Heidegger, cybernetics did indeed make possible communication among the sciences, coordinating physics research, information sciences, psychology, sociology, economics, etc. into the project of control.[5]

2.2 Adapting to the Light

This cybernetic synthesis gives us the theoretical and instrumental background of the exhaustion of subjectivity. However, we have yet to account for how it may take place. The first step is attending to the fact that we have a history; we are molded by what we do and what happens to us. We are plastic, in the sense worked out by

[5] My description of the convergence between behaviorism, cybernetics, neoliberalism, and digital platforms is particularly condensed as it has been recounted in detail elsewhere, especially in Zuboff 2019, Han 2017and 2020, and Gamez 2025.

Malabou. In short, we adapt. The next step is understanding the particular situation to which we are adapting today, knowingly and unknowingly. This process is undertaken by a constant, ever-modulating process of being made into data, a constant telling of the truths about oneself in each moment, which are then correlated with masses of data. It is a turning of every momentary shift of attention, every little spark of desire, each micro-behavior into data, and so turning it into a truth about oneself to be recorded. There is a radical increase in the amount of traces we leave and in the precision of which those traces are analyzed.

In this way, surveillance capitalism is the perfection of Christian confession. Stoneman and Reeves 2024 provide a compelling account of the connection between confession, the contemporary obligation of digital self-expression, and the related damage to subjectivity. The similarity between the confessional booth and digital interaction is jarring in both its low-resolution form of communication and its algorithmic construction. They point out that in the confessional booth, sight is obscured. With body language and facial expressions blocked, the possibility of touch removed, and so on, communication is then reduced to verbal exchange heavily structured by the near algorithmic ritual of confession. This constructs a pseudo-anonymity: we feel a degree of privacy which makes telling truths about oneself easier, but we are actually identifiable and our confessions are retained in memory. Like data, sins are categorized and quantified ("since my last confession, I have had impure thoughts 5 times"); repentance is operationalized ("say 3 Hail Mary's and 2 Our Father's and your sins will be forgiven"), (Stoneman and Reeves 2024, 60–1). While there are an enormous number of our traces recorded at any given moment, for them to be useful, they must all be minute, simplified, classified, and quantified. Our movement, voice, attention, etc. may be recorded simultaneously, but to become data, they must be dissected into individual modalities the way the confessional both restricts the interaction to voice. Citing Han 2021, Stoneman and Reeves describe how the enclosed renunciation of the self of the confessional booth transforms into the "pornographic self-presentation"[6] of social media and digital platforms generally.

What threads these two technologies, social media and confession, together is their offer of absolution, the "cleansing process of transparent exposure," (Stoneman and Reeves 2024, 58). Insofar as platforms stoke the drive towards absolution, they accelerate the confessional destruction of interiority. Authenticity and introspection are often believed to be the special province of interiority, but they are really processes of its destruction. As Foucault analyzes, confession works on the Christian logic of emptying out the self to subject one's desires and beliefs to the scrutiny of others. It clears out the self to make room for truth. This is because confessing is a discursive act where the speaker is also the subject of the statement. Deuteronomy 15:9 tells you to "pay heed to yourself, lest there be a hidden word in your heart." The privacy of the mind is a potential refuge for sin so you must attend to yourself

[6] By pornographic, Han means expressions that lose hermeneutic depth or meaning beyond their surface, which is equivalent to becoming transparent, (Han 2021, 33–4). I quote it here, but do not retain the usage as I find it demeaning to honest pornography.

and submit anything internal to external judgment. Take for example the moralizing dynamics of Twitter where posting becomes the continual subjection of oneself to moral evaluation and hunting out the sins of others. Nothing forces one to post, especially anything private, and yet.

Christian confession demands completeness, but also for it to be difficult on the part of the speaker. There must be at least a pantomime of struggle to express the overcoming of a moral difficulty (Foucault 1990, 61). This latter condition is inverted in the case of surveillance capitalism, which marks an important transformation. Confession becomes constant and easy, and thereby increasingly involuntary even as we desire it ever more. The drive for absolution becomes the background radiation of sociality. Han 2017 sees the transparent society as resulting from a continual voluntary self-exposure that empties out interiority (9). In doing so, we can access what Baudrillard calls the ecstasy of communication ever more easily. The ecstasy of communication is "a phenomenon 'when every-thing becomes immediately transparent, visible, exposed in the raw and inexorable light of information and communication,' when the subject 'becomes a pure screen, a pure absorption, and re-absorption surface of the influent networks'" (Baudrillard 2012, 26), (Stoneman and Reeves 2024, 67). This is to say nothing of the market incentives that propel this continual confession: everyone is their own brand, and (the illusion of) authenticity pays.

Stoneman and Reeves claim that in some sense, the confessional priest is removed by the algorithm, (Stoneman and Reeves 2024, 69). I disagree; rather, the priest is distributed. The mass of observers (interspersed with corporations and governments) all become micro-priests, as if the pastoral vocation were operationalized into its component actions which were then posted separately to Mechanical Turk. We are all deputized clergy, rooting out secrets from the interiority of others, demanding inconsistent penances, reminding others how they are falling short of various salvations, and policing the boundaries of our ever more amorphous communities. As the priestly vocation always demands, this process is a mortification of desire. Once we are reduced to the surface of pure, data-ified behavior, what we want can then be generated and told to us via correlation with masses of data. We are given what we want without the trouble of ever having wanted it. Emptied of depth, organizing human desire is offloaded onto advertising departments, as hideous a transmigration of the soul as has ever taken place. "Marketing has become… the "soul" of the corporation. We are taught that corporations have a soul, which is the most terrifying news in the world." (Deleuze 1992, 6). This emptying out of desire onto the corporation is then tantamount to the end of the subject; were it to be completed, it would be the fulfillment of Arendt's worry.

Worse yet, our voluntary self-exposure via social media posts and the like is a fraction of the data involuntarily collected from us. Data collection is ubiquitous in all digital communication. How long we linger when scrolling is recorded, as is whatever is said near a voice activated device. Data collection extends confession but overturns the difficulty, both nurturing its voluntary forms and massively increasing its involuntary forms. In the west at least, it takes a population cultivated by the practice of confession over centuries, a population shaped to expose itself to judgment and then makes the process easy. In fore-fronting voluntary self-exposure at the expense of the involuntary, Han sees the danger as the possible hijacking of subjects, whereas we are concerned with their exhaustion. Because whatever we attend to online is surveilled, the commodification of attention is an imposition to continually tell the truth about ourselves. And how boring our selves have become in the process. This, of course, makes us more cruel as well; without an ego to discipline (or at least an ego interesting enough to deviate from social conventions), the super-ego turns its drive to punish non-conformity outward.

For all that is unprecedented in this situation, it is comforting to find hints that our cultural memories contain some resources for thinking through it. We will need new concepts and novel syntheses, but we should also use the unprecedented to return to our intellectual inheritances and allow them to show themselves equally anew. Gamez (2025, 113) does just this when he reminds us that Hermes is both the god of communication and a psychopomp who leads our souls to Charon. The great steersman (κυβερνήτης, kybernētēs, the etymological source of 'cybernetics'), Charon, takes his fee and ferries us to the afterlife. Perhaps in future retellings of this myth, the river Styx will fade from prominence, overshadowed by Lethe, the river of forgetfulness. The afterlife awaiting these souls is not Christian heaven or hell, but that of the Homeric pale shades, cognitively diminished versions of their former selves who can barely think, remember, or pay attention. And so, via the apotheoses of communication and control, we are steered through forgetfulness towards an unending existence as pale copies of our former selves.

All for the sake of a few coins.

2.3 Psychic and Social Individuation

I expect some readers will find the idea of the exhaustion of subjectivity, the loss of the meaningful interiority of the human mind, and the offloading of desire to be hard to believe. They may think that interiority is simply a part of the nature of humans as such. Or that the idea of control diminishing our cognitive abilities to predict and modify behavior is solely in the realm of paranoid fantasies and conspiracy theorists. If so, then we are attributing to tech corporations an absurd level of power. Sadowski 2025 raises this concern when he quotes Cory Doctorow critiquing Zuboff as accusing big tech companies of having sorcerous levels of mind control (Sadowski 2025, 33–4). Zuboff's criticisms, he claims, actually function as hype for the advertising services big tech sells to its business customers. What advertiser wouldn't pay for mind control?[7] This is a criticism worth heeding as both the current capacity of these tactics is well below the level of prediction and intervention often attributed to

[7] We should note here that Zuboff's account often depicts the connection between tech platforms and advertisers as unantagonistic. However, as seen in the phenomenon of platform decay, the rent seeking of platforms often puts them at odds with and leads them to offer worse services to their

them (we are not "remote controlled subjects"), and the limits of this lineage of technological development are not yet known. Yet Sadowski himself says later that "a world made into data is a world without subjects," (Sadowski 2025, 86).

Nothing in the exhaustion of subjectivity thesis relies on sorcery. Rather it merely takes seriously that our behavior is moldable by our material and informational environments, and that the common experience of losing hours on an app we don't remember opening or finding an amazon package in front of our door that we do not remember ordering is not the furthest possible limit of this (not to mention that the frequency of the former at least is already a great harm to our well-being and socio-political capacities). Here, there are two immediate responses to those skeptical of this concern.

First, whether possible or not, the exhaustion of subjectivity marks a tendency and fantasy towards which powerful actors employing the techniques of control move. If the exhaustion of subjectivity is a mere fantasy guiding much design and innovation of information and communication technologies, this has the potential for enormous harm even if it is not realizable as such. Biopolitics, for example, maintains the horrific fantasy of the purified nation; this is a fantasy in that racial, ethnic, or any other form of purity is always some combination of nonsense and impossibility. "Pure blood" is a fantasy, but it is a fantasy with a world historical body count that continues to increase.[8] This fantasy is a unique danger that marks a virtual attractor within the state space of biopolitics' diagram of power. If it cannot be actualized, then the exhaustion of subjectivity is at least the same sort of fantasy for control. Whether or not it is possible, it is being engineered towards. For example, Farahany 2023 lays out a range of current brain-computer interfaces and related projects, many of which either contain the possibility of or are actively engineering for the control of individuals through the disruption or prediction and modification of behavior below the level of conscious choice and consent of the subject.

Second, while it is an open question whether the exhaustion of subjectivity is in fact realizable or what forms it could take if possible, it is more possible than the purity fantasies of biopolitics. Racial or national purity has never existed, nor can it exist. Interiority, on the other hand, has a birth date, and everything with a birth date has a death date. At some point in evolutionary history, interiority arose. Whether the human organism can survive it, or it can survive the human organism, I do not know.[9] This knowledge of the contingency of interiority coupled with our lack of a complete scientific theory of the generation of consciousness gives us reason to treat control as potentially more than a dangerous fantasy. This, however, opens the question of what account of interiority could we work with to give shape to our investigation?

---

advertiser customers. This antagonistic relationship may slow the development of control to some degree.

[8] See also Kishi 2024 on the biopoilitical fantasy of race suicide and its sociopolitical effects.

[9] See Lyotard's "A Postmodern Fable."

We have no complete neuro-bio-chemical-socio-informational science of what gives rise to subjectivity. Yet, to understand the problem we are dealing with, we need some account of the genesis of interiority and how it may be exhausted. Having this account will give us a sense of what to pay attention to and where points of intervention are possible. Looking for such an account is to ask after the nature of subjectivity in a way that can make sense of the situation in which we find ourselves. Many common views fail immediately because they cannot do so. If an account of the interiority of the mind implies the impossibility of its exhaustion or that it never existed in the first place, it is not a helpful account for our project. For example, the notion of a soul has a difficult time here because if an immortal soul is the source of subjectivity, there is little sense to be made in how we can be shaped, remolded, and potentially undone by external stimulus. There should be, rather, some untouchable inner core that is beyond the reach of worldly forces. Similarly, a no-self doctrine like that common in Buddhism cannot explain the loss of interiority or the resultant harm thereof because there is nothing to lose in the first place (or its loss or recognition of its never having existed is in fact a goal). This is especially an issue for the Buddhist who has to somehow distinguish this exhaustion from the achievement of nirvana.

The failure of these two views is instructive. An immortal soul and no-self correspond, respectively, to the hylomorphic and the substance/reductionist views that Simondon opposes in his work on individuation. Simondon asks how the generation of subjectivity, or psychic individuation, is possible. What are the conditions through which a mind comes to be able to separate itself from its environment? Individuation generally is an account of onto-genesis, that is, of how individual beings come to be. He argues that to make sense of onto-genesis, some alternative to the hylomorphic view (which smuggles in the pre-existence of the thing by that of its form) and the reductionist view (in which the individual is only ever epiphenomenal or illusory) is required.

The most basic form of onto-genesis is physical individuation. Simondon understands this process on the model of crystallization, where a supersaturated liquid has a condensation nucleus introduced into it, thereby producing a crystal structure that propagates the structure from the nucleus. What is required for physical individuation is an area outside of equilibrium with its wider milieu (i.e., it is not in its lowest energy state), energy which is available for structuration, and the introduction of a condensation nucleus. This results in a metastable structure, meaning something that maintains temporary cohesion and self-identity despite being in non-equilibrium energetic conditions. This structure has a form of "internal resonance," which is a bare minimum of communication between its parts. This resonance or internal communication means, for example, that the energy from an impact on one part of the structure does not affect just that part but gets dispersed throughout the structure. It is here that we can say a new individual has formed; it has become more than a mere collection of parts or aggregate of unrelated components. Deleuze defends a modified version of this that does not require a condensation nucleus, relying instead on the nature of interactions between intensities, and so serves as a general theory of the formation of structure with or without a prior bit of structure (condensation nucleus) to propagate, (see Deleuze 1994, 246 and 251–4, Ardoline 2024, 96–101, and Voss 2025 for more detail).[10] Deleuze's variation is then more favorable to accounts of metaphysical novelty as something new may be individuated in a way that is not parasitic on prior structure (it is also more empirically adequate in that it can accommodate cases of spontaneous symmetry breaking). Where this difference matters, I will be using Deleuze's version. However, in laying out the account I follow Simondon's explication for its more direct explication.

More complex forms of individuation are then built within physical individuation. Whereas the physical crystal forms all at once, biological individuation becomes possible through a slowing down of this physical process. A biological being maintains not just its structure but its structure-forming operations as an ongoing process. To be alive is to be in an ongoing process of individuation. To do so, it must continually make new energy available to this process (i.e., it must consume). Slowing individuation from a singular event to a continuing process is what allows living beings to maintain themselves as a structure while out of equilibrium with their environment. This also establishes a real distinction between inside and outside. A quartz crystal is, in an important sense, all surface. If its outer layer is removed, what is revealed is a new crystalline surface. If the skin of an animal is removed, however, what is revealed is its inside. The crystal reaches equilibrium with its environment quickly; the living being separates itself to (always temporarily) maintain an energetic difference with its environment. We experience this, for example, in the difference between our body temperature and that of our surroundings.

This production of an inside complexifies the internal resonance of the structure as well, making possible robust information storing, response to stimulus, and a minimal sort of agency. "Beings are autonomous when they themselves regulate their own development and store information, and regulate their actions themselves by means of this information," (Simondon 2020, 210). Autonomy or agency here should not be confused with will (or the more robust Kantian form of autonomy we critique in chapter 4); when Simondon refers to agency, he includes the automatic responses of simple biological organisms like amoeba that can distinguish food from non-food and move towards or away from something based on that distinction. This agency is of an immediate stimulus-response loop. Such autonomy persists as the foundation of more complex biological forms such as humans. Greater information dependence means less individuated and less autonomous. This is good and bad, since less individuation means greater plasticity, but also greater dependency on external things. However, for our concerns with the exhaustion of subjectivity,

[10] It is worth noting that there are examples of crystallization that do not require a condensation nucleus and whose structural orientation is unpredictable prior to the process of crystallization itself. This is more in line with Deleuze's account than Simondon's. It also avoids the critique that even if there is a change in kind between the condensation nucleus and the crystal (e.g., dust is different from ice), Simondon's account does fully not succeed in giving an account of a thing coming into being without a pre-existing form because the condensation nucleus supplies its form *qua* its structure which is then propagated, though the structure changes in kind.

complete information dependence is then complete de-individuation. The forces of surveillance capitalism foster greater information dependence through design choices, providing the first of the mechanisms that exhaust subjectivity.

The biological is not enough for psychic individuation. Insofar as a simple biological being is acting on instinct alone, it is not necessarily aware of itself, nor is it capable of reflection. Simondon's account of psychic individuation then explains how the biological inside gives rise to interiority in the cognitive sense. This account also provides a metaphysical backbone for a central concern of the continental tradition: how is it that particular socio-political situations form particular subjects. This process is called subjectivation. Strictly speaking, I will use "psychic individuation" for the general explanation of the formation of subjects, and "subjectivation" for the specific socio-political conditions of particular formations, but this is a formal separation of a single process into its general and context-specific aspects.[11]

Like the slowing down of structure formation into an on-going process complexifies from physical to biological individuation, the slowing down of stimulus-response is the source of psychic individuation. A time between stimulus and response is produced, an interval is sustained, (Simondon 2020, 177–8, see also Deleuze 1986). Generally, this is in response to a problem of a specific form, one that non-reflexive thought is likely to solve less efficiently, one that is disruptive to stimulus-response loops, and that is not an immediate life or death situation. The slow stalking of prey may be such a problem, as would be which way one should travel to find better food.[12] It is in this interval that reflexive thought is possible. If the interval does not exist, then neither does reflexive thought, only simple biological stimulus and response. And like sustaining biological individuation, sustaining the interval requires energy, though it also requires a milieu in which response can be delayed without critical harm. The maintenance of the interval and all the other structures of conscious thought built upon it require that there be an excess of energy beyond the ground state that can be turned into and replenish those structures. For instance, we can open the interval of reflection ourselves when we have excess cognitive energy and no pressing task, hence why boredom provokes thought and why leisure is necessary for reasoning well abstractly.

A mere wedge of time inserted between stimulus and response may seem too slight to be a condition of something as complex and multifarious as conscious reflection. However, creating a temporal difference where there was not one is an incredibly powerful and generative procedure. Compare, for example, how many complex financial techniques become possible because of debt. These techniques

[11] Many thinkers focus on one or the other, or they collapse this distinction (e.g., Stiegler). I find it useful insofar as we can distinguish between the general structure of subject formation (psychic individuation) and different local processes of forming particular types of subjects (subjectivation).

[12] Note that Simondon is not trying to give an explanation of the evolutionary history of psychic individuation, but rather its ontological conditions. The interval surely requires some biological mutation to brains and the like, but we have no complete, agreed upon theory of what that is, let alone how it evolved. Simondon does not pretend to give such an account.

require little more than introducing an interval into the act of purchasing, between gaining possession of something and paying for it.

These are the sites of intervention: information (in)dependence, energy, problem, and the interval as such. Without all these, cognitive interiority is impossible. Yet stimulus-response, that is, behavior, would remain as long as the organism does. The appearance and maintenance of the interval of reflection is what carves out and nurtures interiority. Interiority lingers and persists beyond the interval; we see this in moments where we experience reflective thought but are unable to intervene in our habits, some bodily automatism takes over even as we look for other options. For example, a soccer player may acquire the habit that when things are dropped in front of them, they break the thing's fall with their foot. If so, at some point they will likely drop something heavy and watch themselves automatically bring their foot from safety to directly under the falling thing. They will recognize the coming pain but be unable to pry stimulus and response apart in time to prevent it. These are moments where we experience interiority without the interval. Interiority, then, may linger. Ultimately however, without the interval, interiority eventually withers and closes.

We should not make it sound as if the interval is a moment of pure reflection, though that may be its most extreme form. Rather, what we may draw on in those moments of reflection are those specific resources, habits, skills of thought, beliefs, cultural background, etc. that one has acquired throughout one's life by repetition, adaptation, and in short, learning. This more specific form of shaping the subject is subjectivation, and it will be our focus for the latter half of this book. It is worth noting here, however, that the first step to playing an active role in our own subjectivation (such an active role we can call self-cultivation) is to protect and nurture the interval, to keep the generic process of psychic individuation going. Now, the more the problem situation unsettles one's habits, beliefs, etc. while the interval remains, the more exploratory thought will likely be.

The final individuation process Simondon describes is social or collective; however, the account here is much less detailed than the previous ones. We will attempt to give a more robust account in Chap. 6 through a synthesis with other work, but it is worth describing here for completeness. The slowing down that is required at the social level is not directly stated by Simondon, but there must be a source of social energy, an information dependence between individuals that produces an internal communication structure of the group, and incompleteness of the psychic and biological individuation processes of the group members. This incompleteness and internal communication of the group that subtends the psychic individuals within it is what Simondon calls the transindividual. While we may formally separate them, psychic and social individuation are intertwined, reciprocally interacting processes in a social species such as humans. We carry along not just our biological and psychic individuations as long as we live, but our social individuations as well.

I note here that this book will focus mainly on psychic, social, and technical individuation (the subjects of Chaps. 5, 6, and 7 respectively), even though biological individuation is fundamentally important (its ceasing is death), something we can become active in (e.g., through practices of exercise, nutrition, DIY hormone replacement therapies, etc.), and is altered by control (with technologies such as real time health monitoring or food science engineering). Not discussing biological individuation also means mostly bracketing environmental concerns. This is doubly unfortunate as I embrace the claim that philosophy of technology is ultimately inseparable from environmental philosophy (see Kaplan 2017). It deserves its own study, though we will only touch on it here as an adjunct to these other forms of individuation. This is also true to a lesser degree of social individuation, which in Chap. 6 is not treated in its entirety, but only in its intertwining with subjectivation.

With this account in hand, let's return to the psychic individual to map out how the exhaustion of subjectivity may take place. The exhaustion of subjectivity does not merely undo particular subjectivations, but challenges psychic individuation itself. If the problems that interiority is produced in response to are resolved externally, the process of psychic individuation breaks down. We glimpse this possibility every time we catch ourselves coming back to self-awareness after losing minutes and hours scrolling through one platform or another. The external resolution of psychical problems is exactly what behavioral prediction aims at. This is what big tech promises when it offers us ever more frictionless experiences. A frictionless world is one where we do not encounter problems because they have been solved for us ahead of time, and so the interval does not open. Similarly, the interval itself can be prevented through distraction. One can encounter a problem that would afford the interval, but in the time it takes to begin reflective thought, another stimulus appears and we respond to that instead, often without reflection. This distraction grabs our attention either before the first stimulus-interval-response loop closes or before the interval even fully opens ("why did I open this browser?").

The asymmetry between grabbing and holding attention, and the replacement of holding attention with grabbing attention has become a mainstay of design in surveillance capitalism. Hayes 2025 points out that to make holding attention predictable would require knowing the near infinite varieties of subjects and interests and to be able to correlate them to the specific individuals who find them interesting, for how long given the intensity of their interest, etc. Grabbing attention, on the other hand, is basic and essentially universal (Hayes 2025, ch2). Those who are bored by a conversation about guns still have their attention grabbed by a gun shot. The shift to grabbing attention at the expense of holding it is a disruption of the interval; it is the engineering of an encounter with a problem that attempts to undermine reflection. And as for energy, we have all experienced being too tired for subjectivity. This is often the state the leads us to reach for the phone and zone out in the first place. Energy is required to maintain a structure such as the self and the social groups to which one belongs. The material conditions of undermining the interval include increasing exhaustion through the gig economy, constant work of email jobs, increasingly necessary overtime, etc.

What about the specificity of subjectivation? Psychic individuation is disrupted by the grabbing of attention, the closing of the interval, etc. But the withering of interiority is slow if this is the only process at work. While grabbing attention is easier and this underpins the ubiquity of distraction, that which holds our attention in surveillance capitalism also harms interiority. What is required in both cases is an understanding of individual desire in order to know what will grab or hold attention. Predicting holding someone's attention requires an enormous amount of information indeed, but this is exactly what big data promises. Here we can recall what was said about confession above alongside these techniques for prediction and disruption. These constant traces and recordings betray our desires in minute detail. If you search for something, you will see it again served to you, but even in lingering over an auto-playing video slightly longer than other content will be noted. You will then be given videos with some of but not all the same qualities. Do you linger here too? If so, what do the two videos have in common? In not, how do they contrast? Was it the fire, the music's genre, the actor's hair color? What do you desire such that you react to stimulus faster than the interval allows? What stimulus teases, extends, or satisfies that desire such that the interval does not open yet other distractions are warded off temporarily?

In collecting such data about individuals, profiles are constructed which predict what prompts, distractions, etc. will work on them. This process of collection is the externalization of organized desire onto the profile. A stimulus is provided alongside an immediate solution, so no thought is required to close the interval, or the desire, the enjoyment of the stimulus itself is drawn out, delaying the opening of the interval. The interval between problem and action, stimulus and response, that would provoke subjectivity is shrunk ever further. The entire process is recorded and analyzed to further reduce the window between prompt and action. Insofar as the goal of surveillance capitalism is the perfect modification of behavior, its success is tantamount to the undoing of both psychic individuation and subjectivation.

This account provides the first half of our response to the skeptic of the exhaustion of subjectivity. These techniques intervene and interfere with individual subjects at the very level of their individuation and subjectivation. This puts them beyond the power of conscious, reflective thought to counter, resist, or often even engage with directly. To complete the response to the skeptic, we must show not only that these processes can be disrupted, but what results from their repeated disruption and the externalization of their normally internal conditions. Here, I will argue that the exhaustion of subjectivity is an extreme case of (but thereby continuous with) deskilling. This will clarify what it means for interiority to wither as I claimed above. Doing so explicates the notion of the subject as itself a habit of habits, or the meta-skill of coping with how to live.

2.4 Deskilling the Subject

How serious is this disruption? Is it merely an interruption that leaves no mark on us outside our immersion in our devices? Disrupting the interval required for reflection from time to time does not prevent its future opening. In some idealized dystopia perhaps we could never escape the constant barrage of attention-grabbing distractions, or a perfected brain-computer interface would make it such that we never encounter a problem that provokes reflection. We are not there now. If not, then should we see this primarily as a future outcome to be avoided rather than an ongoing crisis? To answer this, we need to look for what harms to subjectivity result from the frequent disruption of the interval rather than its complete destruction. I contend that there are two important ways in which this takes place: first, in the offloading of the task of organizing desire, which we have touched on above, and second, in the loss of opportunities to practice the skills of subjectivity. Here both are related to what is known as deskilling.

Deskilling is the offloading of tasks which humans do (often onto some external artifacts) in ways that lead to the atrophy of those skills. This idea originally comes from economic analysis, wherein it describes the process by which the skills and knowledge traditionally held by workers are concentrated into management, offloaded onto machines, or made unnecessary by redesigning workspaces and workflows (Braverman 1998). The job no longer requires workers to develop these skills, nor does it allow them time or opportunity to practice them, and so these skills disappear. In addition to the loss of skills, this drives down wages, reduces bargaining power, leads to greater ease of being replaced, etc. all because the job no longer requires someone with specific capacities. Deskilling of all varieties became a wide-scale economic concern during the Industrial Revolution. Marx gives perhaps the earliest account of the political economy of it in the *Grundrhisse* (alongside potentially the first operational account of the development of artificial intelligence). He describes the new relation with machines:

> Not as with the instrument, which the worker animates and makes into his organ with his skill and strength, and whose handling therefore depends on his virtuosity. Rather, it is the machine which possesses skill and strength in place of the worker, is itself the virtuoso, with a soul of its own in the mechanical laws acting through it… The worker's activity, reduced to a mere abstraction of activity, is determined and regulated on all sides by the movement of the machinery, and not the opposite. The science which compels the inanimate limbs of the machinery by their construction, to act purposefully, as an automaton, does not exist in the worker's consciousness, but rather acts upon him through the machine as an alien power, as the power of the machine itself… Labor appears, rather, merely as a conscious organ, scattered among the individual living workers at numerous points of the mechanical system; subsumed under the total process of the machinery itself, as itself only a link of the system, whose unity exists not in the living workers, but rather in the living (active) machinery, which confronts his individual, insignificant doings as a mighty organism, (Marx 1993, 614–5).

Braverman, in setting out the canonical account of deskilling, picks up from Marx and addresses how this process becomes an explicit engineering project via Taylorism. This is often missed by philosophers of technology interested in deskilling. Such an oversight is no doubt a result of attributing to the internal logic of technological development what is more properly understood in terms of advancing capitalist exploitation, even if management is itself an application of cybernetics. Braverman summarizes the process as having three steps: first, the knowledge and organizational skills of workers are recorded and collected. Second, this information is concentrated in and synthesized by management. Third, management takes over as many organizational tasks as they can, relegating the worker to physical labor. The technique of Taylorism is to divide work into mental and physical labor, then distribute as much as possible the mental labor to management and the physical labor to workers. Its goal is to resolve tensions between worker and capitalist interests in favor of the latter (in particular, the tension over how much work is acceptable for what hourly wage, where the worker benefits from less work done per hour, and the capitalist the reverse).

Couple this with the process of rationalizing labor by reducing the physical skills of the worker to their component movements, and replacing ever more of those movements by machines, and you have the twin forces which continually reduce workers to undifferentiated labor power. That is, by deskilling, concrete labor comes to more resemble the abstract notion of labor power, workers become ever more interchangeable, and work itself becomes ever less a form of self-cultivation. Specialization is not opposed to but rather is a driver of this process as it is the condition for simplifying tasks to the point of automation (Braverman 1998, 295). Alasdair MacIntyre notes that this is an ethical catastrophe as well: "the manager represents in his *character* the obliteration of the distinction between manipulative and nonmanipulative social relations," (MacIntyre 2007, 30).

Our concern will be with the cognitive aspect of deskilling. While this economic aspect, especially labor, is deeply important and is mutating under this new form of power, it is not the focus required for this project. It is necessary to keep in mind however both because they set the context in which this project operates, and because the connection between Taylorism and the production of machine learning algorithms is obvious: collect data from human action, train an algorithm, the algorithm then guides or replaces those human actions. Like Marx says of the worker engaged in mechanized labor, we become merely the "conscious organs" of algorithmic platforms. The expansion of information technologies into every aspect of our lives then raises concerns about cognitive deskilling beyond the rationalized workplace. Perhaps we need a corollary to the Cesaire-Foucault Boomerang that whatever is done to colonized citizens by the colonial government will eventually be done to citizens of the metropole: whatever is done to us at work will eventually be done to us in every aspect of our lives.

The proliferation of information technologies coincides with the offloading of cognitive abilities onto artifacts that then results in the inability to complete certain tasks without the aid of the artifact. Everyone who has moved to a new city and relied on GPS to get around knows this phenomenon. If one does not attempt to navigate their new surroundings without GPS, they are unlikely to be able to do so unaided. Insofar as it is through training our skills of moving through a specific environment that we come to feel at ease in a space, our ability to navigate with GPS but not unaided often does not amount to a feeling of comfort, even if we have lived somewhere for years, (Butler 2025).

Because so much of technological development is the offloading of skills onto artifacts, concerns about cognitive deskilling are perhaps as old as technology itself. The most well-known ancient example is Plato's *Phaedrus*, wherein Socrates tells the myth of Theuth, the Egyptian god who invents writing. In the myth, Theuth offers writing as a gift to a king who rebukes him, saying that writing will actually harm people. If knowledge is written down, the people will have no need to exercise their memory, and so will lose the ability to memorize information (as Platonic epistemology demands Recollection, this is potentially quite damaging), (Plato 1997, 551–3). This example is instructive because the king is wrong; writing down information, especially by hand, is an aid to memory. There is a prevailing belief that cultures without writing have vastly superior memories to those with writing, but this is as much a myth as Theuth himself. If there is any important difference here, it is that non-writing cultures are likely to have better skills of improvisation more diffused throughout its members. Deskilling, then, is not something we can deduce *a priori*, nor is it obvious how the function of the artifact will affect us. Writing most mimics our power of memory, but it does not offload the skill it mimics.

Not all externalizations of abilities onto artifacts are a case of deskilling (like writing), nor is all deskilling necessarily harmful (this is perhaps the main difference from the purely economic discussion as there deskilling is always harmful to the worker). If an artifact is ubiquitous enough, there is usually not much concern for needing to perform the skill unaided. Nor is every skill valuable in its own right. Some are worth offloading so as to free up one's time and cognitive resources for other tasks. And finally, upskilling is made possible in that offloading certain skills onto ubiquitous artifacts means we can take them for granted and build new skills on top of them. Ubiquitous calculators improve the level of mathematical skills the average person can reasonably develop, even if they make one less likely to be able to do quick mental arithmetic. The risk, of course, is that knowing when a technological development will be deskilling in a harmful way or not is also not knowable *a priori*. This is both because the results of offloading or externalizing skills are always experimental, and skill acquisition is mediated by cultural, social, and economic factors as much as technological ones. Braverman, for example, is ultimately pessimistic of the supposed upgrading or reskilling of workers, seeing this as a euphemism for capital to create new, smaller niches where the worker always needs less knowledge and skill to fill (Braverman 1998, 309). While I do not think this insight extends to all technological change by any means, it is necessary to keep in mind that deskilling, especially of the pernicious variety, is only diagnosable in the wider context of how subjects are shaped, how production is managed, and the existence or lack of compensatory institutions.

2.5 Externalization

When answering what makes humans distinct from other animals, tool use is often suggested. This then leads to questions of whether beaver dams, bird nests, and the like count as tool use. The response to which is often then to claim the important detail is not that humans are tool users, it's the complexity of the tool use we are capable of due to social memory. Braverman makes this distinction central to his account of deskilling when he opens his book citing a study of South African weaver birds (Braverman 1998, 32). In this study, six generations of weaver birds were raised in captivity, without contact with previous generations. No skills could be passed from the pre-study populations to the post-study populations. Yet, these later generations built nests in ways essentially indistinguishable from their pre-study ancestors, leading to the claim that such nest building is purely instinctual down to the detail. Where Braverman relies on a claim of an essential difference in kind, between instinct and social learning across generations, more contemporary research on nest building in birds suggests that it too has an aspect of social learning (see Breen et al. 2016). We should presume as much of some other animal tool use. The difference of social mediation of tool use between humans and other animals is then at best a difference in degree.

The human difference is not between tool-using animals and non-tool-using animals, nor social versus non-social learning in tool use, but rather how tools come to function as a form of memory for humans. Bernard Stiegler's entire body of work can be understood as the exploration of this fact. The human has, in addition to both the species "memory" of instinct and the everyday sense of memory as retained individual experiences,[13] a third kind of memory, technical memory. Technical memory is the collection of traces of the past retained in artifacts. It is epiphylogenetic in that it is beyond the individual and carried between generations by forces outside the species itself (that is, by artifacts). Technical memory is formed both as recording in the way that we can read off information from a piece of paper and more elliptically in the sense that a tool is always a concretization of human knowledge. A hammer as tool is also the memory of how to build a hammer (recoverable by examining it and taking it apart) and how to swing it (potentially revealed in experimenting with it if one has not already been taught how), as well as a site of organizing social learning (being taught how to swing the hammer by one who already knows how), (Stiegler 1998, 254–5). Furthermore, technical memory makes possible unique features of human consciousness. It acts as "the prosthesis of consciousness without which there could be no mind, no recall, no memory of a past that one has not personally lived, no culture," (Stiegler 2011, 39). This is what characterizes the human: we are the beings that continually offload or externalize our knowledge and memory onto artifacts which are then accessible to and internalizeable by our later selves and others, augmented and modulated through the nature and capacities of these artifacts (Howells and Moore 2013, 5). This reinternalization produces ever greater depths to human interiority.

It is necessary to note that artifacts are never just the concretization of human knowledge, social practices, and the like. The properties and capacities of the artifact shape and change what is concretized and make new possibilities available. In experimenting with a tool, one may discover new uses for it, new capacities it makes available. This is not generally how we think of memory, but it is also not radically unlike the way we may re-interpret myths passed down through generations (that what is remembered can make possible new discoveries is what lends Platonic

[13] This includes both primary and secondary retention in Husserl's terms. Technical memory for Stiegler plays the role of a tertiary retention as well.

Recollection its phenomenological plausibility despite its metaphysical impossibility). Even personal memory is reproduced more so than recalled. In fact, consistent, faithful, and trustworthy reproduction of information is a feature humans find first in certain recording technologies, not our own cognition. Furthermore, artifacts can be improved, altered, retained, or re-engineered based on principles beyond those of biology and culture, and so have forms of change alien to human change over time.

Deleuze and Guattari call the lineages of artifactual change over time the "machinic phylum;" it is shaped by, but not reducible to human choice. Humans rather play the role of pollinators between machines and provide some of the selection pressures guiding machine evolution. This partial independence of the machinic phylum and the reciprocal evolution it undergoes with and through humans (like certain species of wasps and the plants they pollinate) is what constitute the uniqueness of the human. If humans are more able to change, lose, be liberated from, or even preserve their biological, psychical, and cultural processes, it is because we are essentially intertwined with these alien forms of change. Through the construction of our artificial milieus, the human invents itself, but it does not do so only as it chooses. We are human because we co-evolve with the machinic phylum as much or more so than the lineages that make up our biological ecology.

Humans often like to believe we are like the weaver bird was supposed to be. We tell nationalistic or ethnocentric stories of how our various in-groups are "in our blood," some make up pseudo-science like that cellular memory allows them to recall their ancestors, or myths of ancestor spirits that offer guidance and protect traditions. These are just some of the lies we use to repress our technological nature. Cultures result from the retention of social practices and beliefs in different streams of technical memory, not blood memory. Without technical memory, cultural knowledge, beliefs, and practices such as recipes, languages, rituals, religions, and artistic styles can all be lost in a single generation, given an intense enough catastrophe. The Bible (and related artifacts) continually inspires and preserves the Christian god, not the reverse. We tell ourselves what we take to be sacred and important is protected by ghosts, retained as eternal Forms in the mind of god, latent in the blood, soil, or *geist* of a *volk*. All these illusions are repressions of our dependence on and vulnerability to technical memory.

This is the great insight of Stiegler's analysis of the myth of Epimetheus and Prometheus: what determines humans as neither god nor animal, but as mortals, as those who die, is our dependence on artifacts, (Stiegler 1998, 198). If humans flee from our fear of death and what reminds us of our vulnerability, and protheses are signs of our mortality, then we will continually flee from our dependence on technology into our idealist fantasies. Even the existentialist denial of a human essence or belief that an essence is something we make through our own will is a repression of our prosthetic nature. True, our essence does not eternally precede us, and we do make it through making ourselves, but this happens not through our will or intention, but through our prosthetics, through those artifacts we depend on and the memories they preserve. We noted above the common mistake of philosophers of technology to attribute deskilling to artifacts themselves or their evolution alone. This is one more example of the repression of protheses. It depicts our artifacts as foreign, independent sources of harm, and it directs our ire towards them. This perpetuates the fantasy of a human independent of its protheses, an image of ourselves without our signs of mortality (this confrontation with our mortality is redoubled when we remember that the protheses on which we depend were made by (often anonymous) others on who we then depend, especially if we cannot fix or recreate the protheses ourselves). However, if technical memory is essential to the human, then human cognition exists only through technical supports. To echo Clark and Chalmers, we are natural cyborgs and our minds are always, already extended via artifacts.

Deskilling is a result of social arrangements as much as artifacts. Despite referring to the process of having one's skills and knowledge extracted as proletarianization, Stiegler too often falls into this mistake, attributing this loss to technical change itself (see Stiegler 2010). The factory workers of the early twentieth-century were not deskilled by improved machines alone, but by Taylorism, by management. In fact, the goals of management partially defined what counted as an improvement (though not which improvements were possible). When looking at the cognitive deskilling happening today, we must be careful not to attribute to information technologies what is really the result of monetizing behavioral surplus.[14] We must also be attentive to cognitive deskilling as a form of conduct that acts on others' conduct, that is, as a technique of power *qua* control.

2.6 Offloading Subjectivity

What then is the relationship between deskilling and the exhaustion of subjectivity, especially given that the human subject is the result of externalizing our cognition into artifacts? As we saw with individuation, the subject is in some sense the ability to respond to and cope with various problems that we encounter. The subject is neither a soul nor an illusion; it is a habit. It is the habit that results from the active engagement with one's own processes of individuation. It is the (always limited) ability to and ways in which one organizes one's potential actions, reactions, and the like. The subject, in some deep sense, is the ability to be active in one's own self-cultivation. In other words, the subject is itself a habit of habits, or a meta-skill. This recognition connects these ideas and Deleuze and Guattari's work to the virtue ethics tradition more generally. The subject, or character for the virtue ethicist, is a structure of various skills and habits one cultivates throughout their life. Aristotle's term *hexis*, which is generally translated as habit, means a long-term disposition that is robust to change. Habits, on this understanding, are not just repeated, automatic actions, but also the source of the integrity of the subject. Habits cohere into a self.

These terms (self, subject, character) I broadly treat as synonyms; what is important is that we understand the role they play. One's character is simply one more habit that relates and organizes one's other habits. Is a human limited to having one

[14] See also Ardoline and Lenzo 2025 for how cognitive deskilling under surveillance capitalism works in the case of platform decay.

character? Of course not. One is organized in different ways at different times and in different places. However, the more one takes up the active project of shaping a life, the more self-cultivation tends towards a single character. This may be a grave mistake. It is at least a double-edged sword; in some cases, it is advantageous to be one, and in others to be many.

This functional definition of character makes it a subset of Deleuze and Guattari's concept of the body-without-organs (BwO). A BwO is a whole that organizes its parts but sits alongside them as one more part among others. It does not receive a special privilege for playing this organizing role, and it may often be in the interest of the parts to throw off one organization and cultivate another or go their separate ways. Anyone who has ever embraced a conversion experience or decided that they should become a different person understands this (and here the term "any*one*" is misleading insofar as what is being rejected is their character, the very thing which denotes their psychological unity). The BwO is more general in that everything that has some unity has one, whereas virtue ethicists generally limit character to the human alone. I will switch between the two, but it should always be kept in mind that one is a subset of the other. Thinking of character as a BwO will be useful to us as it does not limit our thought to the unaided or technological unextended human (a fantasy if ever there was one), which is often a pitfall of classical accounts of character, but we will return to this in Chap. 5. To further understand the concern over the exhaustion of subjectivity, we can make use of this connection with virtue ethics.

The main claims of virtue ethics are that ethics is a project of cultivating one's self and social groups, and that this cultivation should aim to build capacities that make flourishing more likely (i.e., virtues).[15] If the subject is simply the organizational skill of exploring the connection between one's other habits, skills, etc. and holding them together, then the exhaustion of subjectivity amounts to the possibility of losing the skills of self-cultivation. The erosion of the subject happens by the continual externalizing of the skills one would have to practice to shape oneself well. Surveillance capitalism, then, is the Taylorism of the subject. This means the exhaustion of subjectivity is not only a case of cognitive deskilling but is also what Shannon Vallor calls moral deskilling (Vallor 2015). What we are likely to do, what we are able to do, our emotional reactions, and how we make sense of our practical goings about the world are all largely the result of our habits. What habits we do and do not develop alter even our experience in addition to altering the sorts of actions we are able to take. On Vallor's (and Aristotle's) account, having a virtue means not just having a certain good capacity, but also understanding and being able to explain why a certain action that follows from that virtue in a given situation is good. This amounts to *phronesis*, practical wisdom. The cognitive aspect here clarifies that by habits, we are not talking about mere automatism. Habits are always more than behavior. The notion of habit used here is closer to the embodied contextual intelligence as described by the 4E approach to cognition (Hutto and Robinson 2020).

[15] If there is a difference with classical virtue ethics we must keep in mind here, it is that flourishing is not something we could discover by attending to human nature; it can only be defined provisionally and discovered experimentally.

This cognitive aspect reinforces the fact that shaping one's habits is shaping one's subjectivity. Taking up this shaping as a project is to attempt to become active in one's own subjectivation. To form virtues, we need to have the opportunity to exercise various skills. This is because 1) exercising relevant skills or capacities repeatedly is what forms habits and 2) in the exercising of those skills, one can learn the necessary experiential knowledge required for practical wisdom. Insofar as these skills are conditions of the formation of virtues, they themselves are of ethical concern.[16] Vallor gives several examples of such deskilling. Consider the case where we replace all soldiers with killer robots, all pilots with automated drones, etc. Such a military would have no opportunity for humans to exercise the skills required to form the virtue of courage, or any relevant martial virtues. They would be soldiers and an army in name only. Furthermore, because practical reasoning requires relevant virtues, no human in such a military would be competent to oversee, evaluate orders given to, judge, or make wise decisions about their autonomous weapons. Vallor notes that because they do not have the opportunity to form the proper virtues, drone pilots for example may be exposed to greater risk of trauma. The drone pilot is asked to do a task in a way that does not form the proper habits, and without those habits, they are more vulnerable to the harms those tasks entail, (Vallor 2015, 116, see however Chamayou 2015, 106–13). One could note here also how the Taylorist production of a management class as opposed to the promotion of workers to management roles is the creation of a class who lack the virtues required to understand what they manage (at least as anything beyond sheer profit maximization and manipulative sociality).

A soldier's courage is a particular virtue, needed for a particular pursuit, that is trained through particular activities. However, some skills are general in that they are both relevant to a wide range of virtues one can form and to the very process of active self-cultivation itself. It is these skills that we should be particularly concerned with regarding the exhaustion of subjectivity. High among them is the skill of paying attention. As Vallor points out, attention is fundamental to ethical comportment.

> A person who cannot be counted on to pay attention when you tell her about the recent death of your closest friend, or who is unable to stay focused on the grave and imminent danger to which you're trying to alert her… is not someone who can be said to be virtuous. This is true even of a person who makes a sincere effort to pay attention to her social environment but who has unwittingly lost the cognitive ability to succeed in this task, (Vallor 2015, 117).

---

[16] Above I used habit and skill relatively interchangeably, but this argument of Vallor's requires distinguishing them. This distinction is common in virtue ethics, but I am skeptical that it carves nature at the joints. Skills are understood as the active use of a capacity, whereas habits are the character traits that result from the repeated use of certain skills in certain ways. However, skills are required to form habits and habits determine what skills we have and how they are exercised. So given the mutual dependence of skills and habits, I treat them as referring to the same thing, though skill has the connotation of capacities whereas habit of structures. In Deleuzian terms, this combination of structural features and capacities is the virtual aspect of a thing, and the mutual interaction of these features is the process of differentiation.

Attention, then, is a fundamental ethical skill. It is required to employ what virtues we have and to attend to situations such that we can improve our habits or form new ones. Our self-shaping projects also require that we be able to give them proper attention. We have already noted that attention is under particular threat by control, but here we can give a more detailed exploration of this threat as a form of deskilling. For example, many products of surveillance capitalism make use of algorithmic recommendation systems. Shuster and Lazar 2025 extend Vallor's work and argue that these systems themselves pose a danger to attention. Digital platforms use these algorithms to rank and decide what information is displayed on a platform and in what order. The primary goal of these algorithms on many platforms is not to improve quality of information but to increase engagement. Increasing engagement here generally means increasing the consumption of content. Often this content provokes a strong emotional response. These strong emotional responses reward the user ("reward" in the sense of classical conditioning). Here we see engineering that attempts to avoid the interval: stimulus-response-reward while minimizing reflection as reflection would slow the user down and lower engagement. Shuster and Lazar argue that using these algorithms is then a form of reinforcement learning that rewards multitasking (which is already a euphemism for poor attention), distraction, impulsive behavior, and other forms of disrupting attention. This has direct ethical consequences. Certain skills are required to do ethically important actions. You cannot be a good friend if you are unable to pay attention to a loved one enough to notice they are going through a difficult time because you habitually distract yourself with cat videos, political rage-bait, and the like. This is regardless of whether you value and intend to be a good friend. Therefore, the disruption of one's ability to pay attention, or more specifically, the long-term, repeated damage to their ability to practice the skill of paying attention, is a harm to one's character.

Not all moral deskilling involves offloading. In moral deskilling, a skill may be lost as a result of losing the opportunity to exercise it (and so being less likely or unable to form related habits). The device does not pay attention for us; it disrupts our ability to practice attention. There are of course analogous economic cases where a skill is lost because an industry or manufacturing process becomes obsolete rather than being directly offloaded onto a machine or management, but this is not the focus in discussions of economic deskilling. This difference is enlightening as it shows an important distinction between the economic and the moral cases. What matters in the case of economic deskilling is usually not the particular skills, but whether workers are disempowered by the deskilling.[17] In the ethical case, the particular skills matter as the repeated exercise of the right skills forms virtues, and so are themselves good in some sense. This recapitulates Aristotle's distinction between *techne* and *phronesis*. The skills concerned in the economic case are largely instrumental to some further goal; they are not good in themselves but good insofar as they produce something good. The practical knowledge related to these processes is *techne*.

[17] The exception here would be if there are general skills whose loss is the source of at least some of the alienation of modern labor. A case could be made that the organizational skills Taylorism offloads are of this sort as workers become more alienated the less control they have over their work.

In the ethical case, the exercising of the right skill in the right way at the right time is itself good, both because it is the right action and because it shapes one's character towards virtue. The relevant practical knowledge here is phronesis. However, we should note that this is a formal distinction, not a real one, as ethical and economic deskilling overlap to the degree that the sort of information technologies discussed become tools of managing workers, continuing the Taylorist project. Amazon warehouses and similar businesses attempt to wring out every last bit of labor power from their workers without the worker gaining any skills or work habits in the process, (see Chap. 3 of Crawford 2021). This keeps the worker maximally exchangeable with other workers and makes sure they have nothing to withhold or take to a competitor. While formally distinct, this active attempt to undercut the cultivation of skilled workers through information technologies is continuous with the harm done to the capacity for self-cultivation via these technologies generally.

This is in many ways anticipated by Borgmann 1987 in his critique of the device paradigm. He contrasts a device with a thing. An artifact can be, or be encountered as, either a thing or device depending on the demands they make on us and the artifact's relation to other objects and practices. Things require a certain effort to engage with. Using a thing builds skills. Things also require meaningful connection to other artifacts. The result of interacting with a thing usually produces something we need to then engage with another thing. Engaging with the axe produces firewood that I then need to stack, cover from rain, let dry, etc. before I then use it to engage with the fireplace. Swinging the axe takes effort and so builds skill, absorbs the swinger into the activity, and promotes care for itself and the other objects required to achieve the goal of staying warm for the night.

Devices, in Borgmann's terminology, disburden us. They are easy to use; they do not require much background knowledge or for us to understand how the device relates to other objects or its environment. Devices usually produce something that is immediately consumed, which, in idiosyncratic usage, he calls a "commodity." (Borgmann 1987). There is nothing inherently wrong with a device, but their proliferation, and the way they crowd out things from our lives, Borgmann argues, will make us worse people. First, this is because disburdening essentially leads to deskilling in that we lose the opportunities to expend effort in focal practices, the practices which would both shape us and give meaning to our lives and worlds. Second, because devices do not foster connection with other objects, and (in a Heideggerian vein) our worlds are made by the various connections of concern between things, we and our worlds become ever more fragmented.

Borgmann extends this specifically to information and communication technologies in his account of what he calls "hyperintelligence," which is essentially the interconnected mass of those technologies.

> Hyperintelligence disburdens us from having to remember either the immediacies of schedules, tasks, and appointments or the expanses of history, science, scholarship, literature, languages, or whatever else. A person might be stricken with amnesia while asleep. But as long as, on waking up, one remembers one's name and how to log-on to hyperintelligence, one will be guided from appointment to appointment and furnished with summarizes of what needs done. One can search out in minutes whatever information is needed and give it the appropriate form in a few minutes more.—Borgmann 1993, 107

In this, Borgmann gives us a depiction of the exhaustion of subjectivity *avant la lettre*. One may lose one's memories, one's interiority, and yet with basic access to their smart device and the internet, function well enough in the everyday world.

However, given the account of psychic individuation above, we can see some issues with Borgmann's account. First, Borgmann valorizes effort for effort's sake. Disburdenment is itself harmful in his diagnosis. Devices make things easy whereas things require a level of exertion, and it is this that he takes as the fundamental difference between the two. In this, his view maintains a certain vestige of the Protestant work ethic, which is made explicit in his valorization of work at the expense of leisure, which he takes to be passive and frivolous (Borgmann 1993, 43–4). This is the source of much of his reactionary attitude towards technology and (a not uncritical) nostalgia for American frontier life. In the account of deskilling and its relation to ethics above, there are three differences. One, deskilling is not inherently negative; augmentation and upskilling through offloading are possible. Nor is increasing leisure by disburdening an inherently harmful force, especially because energy is required to maintain interiority. We must keep this in mind even as we diagnose the immense potential harms of control. Two, the specific skills practiced matter as much, if not more, than the mere fact of exerting effort. Due to its status as a fundamental skill, the cultivation of attention is more important here than the supposed inherent good of a hard day's work of chopping firewood. Third, effort for effort's sake is opposed to a central feature of virtue ethics across traditions: the point of repeated practice is to live well with ease, not with continual will and exertion.

Our second major difference with Borgmann is that for him, absorption into an effortful action is seen as a primary good. Insofar as subjectivity is undermined not merely by making tasks easier, but by designing information technologies such that they undermine the conditions of subjectivity, the solution is obviously not to reburden our lives. While it made sending a letter easier, I doubt anyone working an office job in the contemporary world encounters email as a disburdening. Rather, we must proliferate, protect, and nurture the conditions of interiority and the opportunities to practice certain skills, especially those fundamental to cultivation such as attention. This is precisely the nature of leisure, those semi-languorous activities that bracket pressing problems to instead dedicate energy to leaving the interval open. This often means breaking absorption. For example, a class of apps exist to interrupt the user from another app that they find too absorbing, say by giving you a notification that you have been on Tiktok for 15 min and a prompt to ask if you want to continue. This is useful because such an interruption is not an attention-grabbing distraction but the opening of the interval of reflection. These apps are often misunderstood or misused as productivity enhancers; they are better used to protect leisure.

These two criticisms show that the device paradigm does not quite capture the contemporary world of information technologies. Borgmann distinguishes the axe as a thing from the thermostat as a device because the latter does not highlight our connection to other things in its use, does not require effort and so does not build skills, and does not absorb us. An addicting social media app, on the other hand, operates by connecting us with other things (often too fast to appreciate), engages our effort in ways that often do not build skills, and absorb us so deeply into them that we regularly lose hours a week on them unintentionally. In this way, the worst examples of these technologies do not have the ease of devices, nor do they provide the cultivation of things, yet they often promise the former and mimic the latter. One feature of control societies that Deleuze notes is that continual training and retraining have replaced education (Deleuze 1997, 179). A particularly perverse form of this has developed wherein various apps promise manageable practice in certain skills, but the bite-sized practice sessions do not aggregate into skill acquisition. How many people have multi-year streaks in a language on Duolingo, but could not hold a basic conversation with a fluent speaker?

In this way, Borgmann's account is ill fit to the age of the smartphone. Here an equivocation gets in the way: in colloquial terms, the smartphone is the device *par excellence*, the one tool that does seemingly everything. In Borgmann's terms though, the smartphone is not a device. The smartphone does not provide a single commodity but instead enmeshes us in an enormous web of relations to other things, it absorbs us into its continual use, and it does not disburden us in the sense of lightening our exertion even if it makes many tasks easier. While it may provide access to many devices (in the form of certain apps), in Borgmann's terminology, the smartphone is a thing, and a particularly powerful one at that. Yet it facilitates and accelerates the harm to the subject that Borgmann attributes to the device paradigm. The flaw that leads to this ill fit is, as I argued above, that Borgmann maintains an unsupported belief in the inherent goodness of work as strenuous effort over ease, and in absorption into a task. This is also at odds with virtue ethics in that the virtuous person is not one who struggles with temptation or other vices but then does the good by overcoming them; the virtuous person does what is good, at the right time, in the right way, etc. as if second nature, that is, with little effort.

2.7 Conclusion

Opposed to the Borgmannian worry of deskilling as disburdenment, I propose the source of the exhaustion of subjectivity and the unique harms that result from it is instead the Stieglerian offloading of desire onto technical memory (this is, however, different from Stiegler's own account of disindividuation[18]). This brings together the various pieces of this chapter. Human subjectivity is the product of processes of

[18] Stiegler attributes the disruption of our processes of individuation to ever greater synchronization. He claims consciousnesses are differentiated by being different times, yet we collectively engage with more and more of the same temporal objects (watch the same television, listening to the same music, hear the same news, etc.). Therefore, we come to occupy the same time flows of consciousness, thereby becoming synchronized and less distinct from each other (Stiegler 2009a, 2009b, 43–5). This account fails in a few ways. First, this trend of media consumption has reversed dramatically in the age of algorithmic curation, streaming services, and micro-celebrities, yet individuation is still threatened. Were he right, the breakdown of the monoculture, say in America, should lead to greater individuation and stronger senses of self-identity alongside it. More importantly, it falls afoul of Stiegler's own criticisms of idealism. Somehow synchronization is not materially mediated and our absorption into the same temporal objects shapes us without the copying error one would expect in any form of material retention, synapses included.

psychic and social individuation that are mediated by technical memory. The techniques of surveillance capitalism accelerate to a nigh unfathomable degree the collection of data about us, with the limit point of a continuous process of complete confession. This amounts to a process of deskilling important cognitive abilities, including paying attention, memory, and the like (we should note that this also occurs alongside a process of technical deskilling made possible ever sleeker devices and intuitive user interfaces that then require ever fewer people actually understand how these devices work and how to modify them, see for example, how Gen Z has less understanding of how computers works than Millennials). This nurtures greater informational dependence which as Simondon argues leads to greater de-individuation. The goal of this is to produce ever better prediction and behavior modification interventions. However, we have seen it is less important how effective these predictions and modifications are than the cognitive deskilling effects they have. That is, while they are no doubt impressive in the interventions they make possible, the greatest danger is not that they become omniscient and omnipotent, but that we become ever more predictable and malleable, that our behavior becomes such that interiority and its related concepts become unnecessary theoretical posits.

This would be the exhaustion of subjectivity. Subjectivity is here understood as a skill of skills; the self just is the organization of our various other abilities, habits, beliefs, etc. As more of these pieces of ourselves are offloaded or deskilled, there is less and less of an interiority to our lives. Again, attention and memory are of the utmost importance. Attention atrophies (deskilled but not offloaded) whereas memory is ever more externalized. Attention makes possible care, and memory makes possible prediction, anticipation, and imagination. Care and anticipation, concern for some imagined future, is the very heart of desire. A simple truth unites so many thinkers, those such as Xunzi, Spinoza, Freud, and Nietzsche, whose insights their surrounding cultures can neither reject nor bear and so must preserve while repressing. The truth is this: desire is the material of subjectivity.[19]

Finally, perhaps the skeptic persists that interiority is not truly under threat. We have said that the subject or the self is a collection of habits, or a habit of habits. Similarly, the products of surveillance capitalism are made to be addictive, that is, habit forming. Varoufakis 2024 points out that we are trained in new habits regularly, his example being learning how to engage with voice-activated assistants like Alexa. If so, then is this worry of the exhaustion of subjectivity misplaced? There are two concerns here. First, not every collection of habits cohere into a subject. One is at odds with oneself to some degree, and it is possible that the habits formed from these products are not capable of being integrated into a subject. Second, if, on the other hand, they do indeed cohere into a subject, what sort of subject is it? That is left open, the goal is not to shape anyone in particular, but to produce behavior. Do we have any sense of who we are trying to become in doing so? In the experimental

[19] When this truth is faced by accepted major figures in various intellectual traditions, it is almost always lamented or feared (see Augustine on desire or the Buddha's four noble truths) or desire is defanged (see Mengzi's exclusion from human nature of anything but our "good" desires or the liberal reduction of desire to rational self-interest aimed at efficient satisfaction).

task of forming a life, we are always in a deep sense unsure of who we are becoming, but whatever subject these new habits may form, it is one that is radically disempowered to engage in its own self-shaping. This, then, is the great danger facing us from whatever this new form of power is. The matter, resources, and skills of forming ourselves are under attack. At worst, subjectivity itself is threatened; at best, we are producing subjects with little to no capacity to shape themselves. We are, in short, exhausted.

References

Ardoline, Michael J. 2024. *Deleuze, Mathematics, Metaphysics: Difference and Necessity*. Edinburgh University Press.

Arendt, Hannah. 1998. *The Human Condition*. 2nd ed. The University of Chicago Press.

Baudrillard, Jean. 2012. *The Ecstasy of Communication*. Trans by Jean-Louis Violeau. Semiotexte.

Borgmann, Albert. 1987. *Technology and the Character of Contemporary Life: A Philosophical Inquiry*. The University of Chicago Press.

Borgmann, Albert. 1993. *Crossing the Postmodern Divide*. University of Chicago Press.

Braverman, Harry. 1998. *Labor and Monopoly Capital: The Degradation of Work in the Twentieth Century*. 25th Anniversary ed. Monthly Review Press.

Breen, Alexis J., Lauren M. Guillette, and Susan D. Healy. 2016. What Can Nest-Building Birds Teach Us? *Comparative Cognition & Behavior Reviews* 11:83.

Butler, Michael. 2025. The City, the Highway and the Spatial Level. *Techné: Research in Philosophy and Technology* 29 (1): 69–83.

Chamayou, Grégoire. 2015. *A Theory of the Drone*. The New Press.

Crawford, Kate. 2021. *Atlas of AI: Power, Politics, and the Planetary Costs of Artificial Intelligence*. Yale University Press.

Deleuze, Gilles. 1986. *Cinema 1: The Movement-Image*. Translated by Hugh Tomlinson and Barbara Habberjam. University of Minnesota Press.

Deleuze, Gilles. 1992. Postscript on the Societies of Control. October 59: 3–7.

Deleuze, Gilles. 1994. *Difference and Repetition*. Translated by Paul Patton. Columbia University Press.

Deleuze, Gilles. 1997. *Negotiations 1972–1990*. 1st ed. Translated by Martin Joughin. Columbia University Press.

Farahany, Nita A. 2023. *The Battle for Your Brain: Defending the Right to Think Freely in the Age of Neurotechnology*. St. Martin's Press.

Foucault, Michel. 1990. *The History of Sexuality, Vol. 1: An Introduction*. Reissue ed. Vintage.

Galison, Peter. 1994. The Ontology of the Enemy: Norbert Wiener and the Cybernetic Vision. *Critical Inquiry* 21 (1): 228–266. https://doi.org/10.1086/448747.

Gamez, Patrick. 2023. Inhuman Hermeneutics of the Self: Biopolitics in the Age of Big Data. *Foucault Studies* 34:80–110. https://doi.org/10.22439/fs.i34.6939.

Gamez, Patrick. 2025. *Posthumanism Meets Surveillance Capitalism: How to Delete the Manifest Image*. Palgrave Macmillan.

Han, Byun-Chul. 2017. *Psychopolitics: Neoliberalism and New Technologies of Power*. Verso.

Han, Byun-Chul. 2020. *The Disappearance of Rituals: A Topology of the Present*. Polity.

Han, Byun-Chul. 2021. *Capitalism and the Death Drive*. Polity.

Hayes, Chris. 2025. *The Sirens' Call: How Attention Became the World's Most Endangered Resource*. Penguin Press.

Heidegger, Martin. 2008. In *Basic Writings*, ed. David Farrell Krell, Revised, Expanded ed. Harper Perennial Modern Classics.

Heidegger, Martin. 2013. *The Question Concerning Technology, and Other Essays*. Reissue ed. Harper Perennial Modern Classics.
Howells, Christina, and Gerald Moore, eds. 2013. *Stiegler and Technics*. Edinburgh University Press.
Hutto, Daniel D., and Ian Robertson. 2020. Clarifying the Character of Habits: Understanding What and How They Explain. In *Habits: Pragmatist Approaches from Cognitive Science, Neuroscience, and Social Theory*, ed. Fausto Caruana and Italo Testa. Cambridge University Press.
Kaplan, David M., ed. 2017. *Philosophy, Technology, and the Environment*. Mit Pr.
Kleinherenbrink, Arjen. 2024. Getting Real about Artifacts: OOO, Post-Phenomenology, and Technology. In *15 Years of Speculative Realism: 2007–2022*, ed. Charlie Johns and Hilan Bensusan. Zer0 Books.
MacIntyre, Alasdair. 2007. *After Virtue: A Study in Moral Theory*. 3rd ed. University of Notre Dame Press.
Marx, Karl. 1993. *Grundrisse: Foundations of the Critique of Political Economy*. Translated by Martin Nicolaus. Penguin Classics.
Mirowski, Philip. 2006. *Machine Dreams: Economics Becomes a Cyborg Science*. Cambridge University Press.
Pasquinelli, Matteo. 2023. *The Eye of the Master: A Social History of Artificial Intelligence*. Verso.
Plato. 1997. In *Plato: Complete Works*, ed. John M. Cooper and D. S. Hutchinson. Hackett Publishing.
Reeves, Joshua, and Ethan Stoneman. 2024. From the Confessional Booth to Digital Enclosures: Absolution as Cultural Technique. *Theory, Culture & Society* 41 (4): 57–73. https://doi.org/10.1177/02632764241233227.
Sadowski, Jathan. 2025. *Mechanic and the Luddite: A Ruthless Criticism of Technology and Capitalism*. University of California Press.
Simondon, Gilbert. 2020. *Individuation in Light of Notions of Form and Information*. Translated by Taylor Adkins. Vol. 1. University of Minnesota Press.
Stiegler, Bernard. 1998. *Technics and Time, 1: The Fault of Epimetheus*. 1st ed. Stanford University Press.
Stiegler, Bernard. 2009a. *Acting Out*. Translated by David Barison and Daniel Ross. Stanford University Press.
Stiegler, Bernard. 2009b. *Technics and Time, 2: Disorientation*. Translated by Stephen Barker. Stanford University Press.
Stiegler, Bernard. 2010. *For a New Critique of Political Economy*. Translated by Daniel Ross. Polity.
Stiegler, Bernard. 2011. *Technics and Time, 3: Cinematic Time and the Question of Malaise*. Stanford University Press.
Vallor, Shannon. 2015. Moral Deskilling and Upskilling in a New Machine Age: Reflections on the Ambiguous Future of Character. *Philosophy & Technology* 28 (1): 107–124. https://doi.org/10.1007/s13347-014-0156-9.
Varoufakis, Yanis. 2024. *Technofeudalism: What Killed Capitalism*. Melville House.
Verbeek, Peter-Paul. 2005. *What Things Do: Philosophical Reflections on Technology, Agency, and Design*. 1st ed. Penn State University Press.
Voss, Daniela. 2025. 'Singular Without Being Individual': Simondon, Deleuze, and the Problem of Individuation. In *The Deleuzian Mind*, ed. Henry Somers-Hall and Jeffrey A. Bell. Routledge.
Wiener, Norbert. 1971 *Human Use of Human Beings: Cybernetics and Society.* Discus Avon.
Zuboff, Shoshana. 2019. *The Age of Surveillance Capitalism: The Fight for a Human Future at the New Frontier of Power*. PublicAffairs.

Chapter 3
Control Come to Light

3.1 Introduction

In the previous chapter, we examined the conditions of the exhaustion of subjectivity. Following Gamez's account, I maintain that this is a uniquely important phenomenon for diagnosing the forms of power we are seeing arise alongside the proliferation of information technologies. Characterizing this form of power is then the next step in the analysis. In work that engages this question, some thinkers attempt to show that this mutation of power is still within the categories that Foucault laid out, including Gamez's claim that surveillance capitalism is a continuation of biopower. I disagree. I will argue that we better understand the nature of the problem, why it is a problem, and how to strategize against it through the category Deleuze suggested: control.

Deleuze was circumspect enough to realize that when he was writing in the early 1990's, this new form of power had neither fully shown itself nor completely taken shape. This is in some sense still the case. However, in this chapter, I will try to fill out this account both by showing how control builds on but exceeds discipline and biopolitics and give a general characterization of its unique conditions and mechanisms. To do so, I will recount the essential features of biopower that Foucault elucidates (the management of life, focus on individuals and populations, cultivation of discipline, and the use of statistics) and showing this contemporary form of power either does not share in these features (it manages access via continual modulation, focuses on dividuals and masses, and undermines discipline) or it builds on them (it repurposes the tools of discipline for behavioral modification and the use of statistics is transformed by linear algebra and programming).

M. J. Ardoline, *Interiority Against Control*, Sophia Studies in Cross-cultural Philosophy of Traditions and Cultures 47,
https://doi.org/10.1007/978-3-032-16245-8_3

3.2 Strata of Power

As much of the work we have from Foucault on biopower is through his lectures, the account is very much a work in progress. It is often noted that he used biopower and biopolitics sometimes interchangeably, and sometimes with a distinction. In the latter case, biopower is the wider category of techniques of power, where discipline names their application to individuals and biopolitics names their application to populations (Foucault 1990, 139–41).[1] I will follow the latter usage. Biopower is then a strata of power following sovereign power, that first forms through the accumulation of disciplinary techniques, which later allies with population-level techniques organized around the strategy of fostering life.

A stratum of power is formed by the slow accretion of methods, techniques, devices, norms, categorizations, rituals, and so on, which at some interesting point cohere into a dominant strategy. Strata do not mark sharp breaks with what came before, nor are they "pure." Rather, they are suffused both with the components of previous strata and with anarchic elements of power which have never cohered into a dominant strategy. Biopower maintains the tool of state execution from sovereign power, even if it occasionally repurposes it, for example. This makes the detection and diagnosis of a new dominant strategy particularly fraught. Its components will be found in earlier strata, well before it has taken form, where other historical paths were possible. It is also complicated by the persistence of previous strata (anyone wanting evidence that biopower remains ascendent will find plenty in the current capture of the United States' government's health services by anti-vaccine, pro-eugenics activists). Furthermore, the anarchic elements are always liable to be misidentified as characterizing a new stratum of their own (and in truth, they may eventually do so). This is in addition to the fact that rarely does any group identify with a particular dominant strategy; rather, they mark the implicit background of certain eras, assumed in common between various otherwise opposed groups.

I do not argue that control has fully replaced biopower, nor that control has reached its apex. Even if it is ascendent, exercises and techniques of both sovereign power and biopower will remain in use under control. Rather, we are at a point where control has cohered, and so its nature is becoming knowable; its techniques are becoming distinct and some of its unique effects are being felt. Nothing shows its essence at the moment of its birth. This is partially because essences are historical and partially because nothing has the power to stand on its own at birth; it must then hide for a time. Deleuze says of cinema that it first needed to mimic human

[1]There is a complicated historical element to this categorization. While discipline comes to fall under biopower as its application to individuals, the chronology of Foucault's analyses suggests that discipline arises first as a new form of power succeeding sovereign power. However, once biopolitical techniques take shape, they are then coupled with disciplinary techniques, giving rise to the category of biopower as the combination of the two (Foucault 2003, 249). This arrangement is then a historically contingent one where an independent strategy of power is absorbed and put to use by another, in a way analogous to endosymbiosis.

perception to be embraced and explored. Only once it had become familiar to enough people did it reveal its possibilities of non-human perception, of recording and projecting experiences that no human eye could ever perceive on its own (how much cultural scaffolding was required to allow us to make sense of the montage?). What I argue is that the assembly of forces analyzed in the last chapter is a sign that control is starting to show what only it can do, akin to early movie goers first seeing the sudden cut from a vast wide shot to a close up of an actor's face. It is neither just born nor finished growing, but its diagrammatic essence, its virtual structure, is being shown.

A form of power being dominant does not mean its tactics always work or are even particularly effective. Sovereign power did not deter theft by punishing the crime with death. If anything, an account of sovereign power explains why such a program appears as plausible to those within it, and what effects it actually has that ensure its persistence (e.g., legitimating privatizing land through enclosure). That is, understanding a form of power will, among other things, explain why certain beliefs are taken as true or obvious regardless of their actual truth value. I will not follow the common Foucauldian practice of calling any accepted set of beliefs that serve or are supported by legitimating functions "a knowledge."[2]

Finally, it is worth noting that Deleuze attributes to Foucault himself the insight that a new dominant form of power is being assembled (Deleuze 1997, 177–8). Deleuze does not cite any particular claim or example from Foucault, but such a claim is plausible when, say, reading *The Birth of Biopolitics* and explaining the misleading title. In this lecture series, Foucault says he will first lay out liberalism in order to explore how biopolitics could take shape in such a context. Here the supposed problem is that the rise of liberalism means the concept of the individual has a powerful legitimating role, yet biopolitics is aimed at the population in ways that ignore or violate the supposed liberal primacy of the individual, (Foucault 2010, 317). Instead, the lecture series becomes a history and analysis of neoliberalism's then recent rise to prominence. We saw in the last chapter how that history is central to the history of control. It is not too wild a speculation to think that Foucault sees in this historical mutation of liberalism the first sedimentations of control which had not yet coalesced. Deleuze at least seemed to think so.[3] We will turn now

[2] I take Foucault's power/knowledge account to be a misnomer that leads to several common misuses. Power explains not knowledge as such, but what beliefs become accepted, true or not. To hold this does not mean that we have any guaranteed way of separating truth beliefs from false, just that such a distinction exists. One such problem produced by this misnomer is that the equation of knowledge with power couples with the naïve belief that power is always bad, leading to either the belief that every claim which serves power must really be false (since it is only legitimated by some usually unjust power structure) or that knowledge itself is an oppressive category to be opposed. While the misnomer is Foucault's, he rejects both of these misunderstandings.

[3] Foucault mentions control several times in the lecture, especially page 67, though this seems to me to be at best inspiration for Deleuze's use of the term, not the naming or diagnosis of a new form of power. More likely, it is a post facto equivocation resulting from the commonness of the word, especially in the context of discussions of power.

to looking at several attempts to analyze the forms of power made available through information technologies in terms of biopower.

3.3 Discipline

Discipline (or disciplinary power, or anatomo-power) "centers on the body, producing individualizing effects, and manipulates the body as a source of forces that have to be rendered both useful and docile," (Foucault 2003, 249). It is the aspect of biopower that aims at shaping individuals, and does so such that the result is a subject of a certain type, with certain character traits, dispositions, etc. Foucault explicates it in distinction with sovereign power *qua* the power to kill or let live (Foucault 1995). In this earlier stratum of power, one lives at the pleasure of the sovereign. This power is justified by and given shape through the imposition of law, of which the sovereign is the ultimate judge. The law is concerned merely with whether the act has been carried out by the accused. The trial is undertaken in private; the punishment is a public spectacle, ostentatious and laden with intentional symbolism (e.g., the hand of the regicide that held the knife *must* be punished more than the hand which did not). Like all strata of power, the absoluteness matters less than the particular diagram or constellation of components and how that diagram is transposed throughout society. It is here less important that a knight tasked with killing a random peasant on the fringes of the kingdom may slack off, ride around for a few weeks and return to declare it done with little fear of being caught in his lie, than it is to understand that when transposing from the state to family, a husband becomes the *pater familias*, the sovereign of the household who has the legal right to kill his wife or children.

In disciplinary power, all of this is challenged, though not always reversed. The law is replaced with the norm or the model of the good subject who fulfills a certain role. Institutions and techniques are developed and coordinated to produce this subject. While sovereign power is mainly prohibitive (thou shall not), discipline makes positive prescriptions of lifestyles, subject types, etc. (Foucault 2009, 46). The army aims to produce the good soldier, the school, the good factory worker, etc. Historically, this move from sovereign to discipline follows the rise of collective, regularized factory work (Han 2017, 19–20). As such, central to discipline is the architecture of enclosed, easily surveilled spaces, of which Foucault takes the panopticon to be ideal limit. While well understood, it is worth briefly reiterating why: the panopticon is a prison structure in whose center is placed the guard's station and around which are the cells. The guards can see into any cell at any time; however, the other direction is obscured such that inmates never know if a guard is watching them or not. There are clear norms and rules for behavior, for a good inmate. The mechanism this is supposed to foster is that in never knowing if they are being watched, the inmates must always think and act as if they are, which over time is supposed to lead to these rules being internalized. One constantly watches oneself, becoming their own guard, enforcing the rules on themself, and so comes to conform to the norm. The justification for punishment is now moral improvement and reformation (we should be careful to distinguish justifications from actual goals). This is a dominant diagram in that one can then diagnose full or partial implementations across institutions (the prison is like the hospital is like the classroom is like the factory are all like the prison).

De Laat 2019 argues that machine learning and big data do not alter but rather merely intensify this dynamic. These technologies make possible new forms of targeting individuals, but, he argues, this targeting remains broadly under the scope of discipline and the panopticon. His analysis focuses on how norms and normation remain operative in machine learning, especially in supervised machine learning. Here, big data is used to generate systems that recognize, categorize, predict, and intervene based on norms. Furthermore, the increase in data collection throughout society is essentially extending the panoptic gaze beyond its usual enclosures. The opacity of the panopticon is extended insofar as algorithmic decision making is often black boxed; we believe that the system works, but we do not know how it works. If it serves as decision maker or advisor, we generally cannot ask it for its reasoning or justifications (attaching an LLM like ChatGPT to such a system is just asking it to produce *ad hoc* and *ex post facto* explanations that we have no reason to trust have anything to do with algorithm's operation. Admittedly, this is incredibly human of it). All of this, he argues, only increases the pressure on individuals to conform to norms. "All subjects involved are drawn into a comparative field which, by its very existence, exerts a pressure to conform," (de Laat 2019, 321). Insofar as normation remains, De Laat argues, so remains discipline.

This is all reasonable and correct as far as it goes (it is *surveillance* capitalism, after all), but it misses the important and unique mutations that happen in power's relation to subjects, or the lack thereof, in these new techniques. There are three mutations in particular that make discipline an ill-fit framework for understanding this form of power, all three of which we see in the exhaustion of subjectivity: the abandonment of shaping subjects, the construction of dividuals, and the repurposing of surveillance.

First and foremost, discipline is the shaping of certain types of subjects, yet this new form of power is directly antithetical to the formation of subjects as such. The sorts of interventions this form of power make possible are ones that disrupt the formation of habits; they externalize the interior to make it available to technical memory. Discipline works in the reverse direction through pressuring subjects to internalize norms and social roles. Discipline deepens interiority, even if it does so for the sake of self-policing. The different roles of opacity in discipline and control make this distinction clear. The inmate in their cell is given a role, a list of expected behaviors, a series of societal norms, etc. that they are meant to internalize. The guards who evaluate whether these criteria are being met are hidden from view, but that they are viewing you and what they are looking for is made clear. Here opacity generates the illusion of constant surveillance to facilitate this internalization.

Control does the reverse. Surveillance *is* constant, but is often invisible: tracking cookies, hidden security cameras, programs for scraping social media posts, etc. All these forms of surveillance are either hidden completely or disguise the fact that they are surveillance. To lure out the secret depths of one's interiority, the illusion of not being watched must be cultivated. Because we do not know the where, how, what, or why, this fact remains even if we are all now paranoid, constantly believing everything we are doing is being recorded. We know this generally, and yet we constantly are lured into believing it is not the case. For example, people are often surprised to learn the Captchas are also a form of data mining for machine learning. Early Captchas had users demonstrate their humanity by recognizing sloppy handwriting. This no longer works because those interactions were used as part of the machine learning process that led to the technical marvel that is contemporary Optical Character Recognition software. The very process of proving one's humanity is itself a part of externalization onto technical memory. Discipline equally made use of tasks for the subject, but the goal was to produce certain capacities in a subject; here the subject is surveilled through a task without realizing it in order to externalize that ability, and potentially deskill them (I, for one, cannot read my students' handwriting but OCR can).

Not only is the fact of surveillance itself opaque, the norms at the source of judgments about us are hidden. Social media companies struggle to keep private not just how much they know about their users, but also the categories into which they place them. If these categories are indeed norms (and given that they may be unrecognizable to humans if produced from unsupervised learning, and we are unaware of them, I am not sure some of them are norms in any meaningful sense), we do not know what it is we are asked to conform to. "We are not *disciplined* by this surveillance, nor could we be—we do not even know what people and machines surveilling us want from us, and it could be many different kinds of interests that even contradict each other," (Seecamp and Söffner 2024, 76, see also Koopman 2019, 166).[4] We should point out the capitalist pressure towards this form of surveillance. A corporation doesn't care if you break a norm or not; it cares whether it makes money. It will intervene in whatever way does so. If we do not know what we are to internalize (and our very processes of internalization are disrupted), then in what sense is this form of power disciplining subjects?

We should not rely here on the exhaustion of subjectivity being achievable. Even if we take the more skeptical stance that this phenomenon is a limit case and a guiding fantasy, the relation of these new techniques of power and types of subjects is completely different. Discipline attempts to make behavior predictable by making a certain type of subject, someone docile or trustworthy, who identifies with their social role, etc. That is, it produces individuals of the normed type in order to get predictable action. In control's cybernetic behavioralism, we saw that character or subject type is treated as an unnecessary theoretical posit. Control aims to produce the desired action by targeted intervention, and if a subject is produced at all, the normed type will be produced as accidental residue (see for example, the enormous proliferation of baffling internet subcultures). This is what makes scandals such as

[4] I should note that Seecamp and Söffner take us to be moving into something beyond control societies. I disagree and instead take their diagnosis of this movement to be an account of how control has further been revealed or mutated since Deleuze wrote the original piece.

Cambridge Analytica that much more terrifying even if its initial claims were overblown: before we know what kind of person votes for Trump we know how to make people more likely to do so. The minutia of control makes possible the prediction of action beyond any shape of a type of person; it makes possible the production of consistent actions before the existence of the norm. As Overwijk 2025 analyzes, this becomes all the more clear in the January 6th storming of the U.S. Capital where internet-enabled syncretic conspiracy theories blended with fascist aspirations to produce coordinated action despite the radical incoherence of the various participants (Overwijk 2025, 176). By incoherence, I do not mean that their individual beliefs were merely absurd (though they were), but that there was an enormous variance in what each individual or subgroup believed they were doing and why. Each subgroup had whatever bespoke ideology, conspiracy theories, and mistaken beliefs that they needed to motivate them to this coordinated action.

Second, following from the first, this form of power differs from discipline in that people are tracked, surveilled, and modulated not as individuals, but as dividuals. If types of individuals are produced only accidentally if at all, what is being targeted for intervention? Discipline individualizes a body as an organism with capacities to be shaped (Foucault 2003, 249); what we see in control instead is the dividualization of persons. "Dividual" is a term Deleuze coins but does not adequately define. I hazard a tentative definition here: a dividual is a subdivision of a person or group into particular traits, desires, capacities, or behaviors, which are correlated to other traits, desires, capacities, and behaviors by using the data of others to model likely transformations of those traits, desires, capacities, or behaviors. Patton 2018 describes this admirably

> Dividuals are not whole subjects but partial subjects defined by certain functional aspects identified in relation to particular ends. The same biological person might correspond to a financial capacity to repay a bank loan, as defined by their age, income, lifestyle and existing level of debt; to an intellectual capacity to undertake a given programme of study, as defined by their prior education and levels of achievement; and to a physiological capacity to qualify for a certain insurance product based on their medical history, genetic make-up and lifestyle. Each of these financial, intellectual and physiological dividuals can be tabulated and recorded along with others of the same kind to form a data bank that can be analysed and exploited for commercial, governmental or other ends.—Patton 2018, 195.

These correlations are then used for the sake of predicting and intervening in behavior. The dividual is paired with the mass of data, to establish these correlations and make prediction possible. One is then understood not as a holistic individual, nor even in terms of one's own properties alone, as there is no guarantee that one has all the traits a dividual is correlated with. In fact, it is more important that you do not yet have all these traits, because the goal is to shape behavior. In other words, surveillance capitalism treats you neither as a being with a unique soul, nor as a Humeian bundle of properties. You are a collection of properties you both have and might come to have, without the boundary of a self to distinguish the two. You are modelled as a collection of likely transformations. Note that the ontology implicit in this is neither hylomorphic or reductionist; rather, it works with the same ontology as individuation, at least the level of process. In short, to understand people as dividuals, one grasps them not as fully formed subjects or individuals, but at the level of their constituting processes of individuation. Discipline aims at a finished product, a shaping of human clay that is eventually fired in a kiln to retain its shape and function permanently; control never stops shaping and reshaping the clay.

This leads to a third important difference. Surveillance in this form of power is not used to detect and punish those that break the law or norm; it feeds a system of prediction and intervention that attempts to modulate behavior (towards a norm or otherwise). De Laat recognizes this difference though he sees it as an intensification of discipline: "Primary discipline is put into action as soon as a subject breaks—or appears to be about to break—the norm; the secondary mechanism of discipline based on prediction as an outcome of ML [Machine Learning] may suggest an intervention even before that: as soon as a subject appears likely to be breaking a norm sometime in the future. The former discipline had to wait until norm deviance happens or seems immanent; the latter anticipates it and acts accordingly," (de Laat 2019, 323). However, he does not appreciate how this is opposed to the very notion of disciplinary power. Not only is this form of surveillance often hidden as noted above, it is no longer used to shape a subject, but to intervene directly on behavior.

Given the way strata of power sediment, that there is a case to be made for a continuity with discipline is not surprising. Furthermore, since on Foucault's account there are not clean breaks nor does one dominant strategy replace the previous ones completely, we should then expect new examples and mutations within sovereign and biopower. However, this also presents a danger in seeing these changes as evidence of a continued dominance of a strategy that has passed or that a new one has not formed. What has happened here is that the components of discipline have been recognized (surveillance, occasionally norms), but not how they have been reconfigured and repurposed into a new diagram. It is as if an archeologist found tools of an era earlier than expected at a dig site and concluded they remained in their traditional use longer than previously believed, but really had stumbled across the remnants of an ancient society's museum.

The pivot point between discipline and control is the brain. One can almost feel the turn in Stoneman and Reeves 2025's recounting of the aims of the failed MK Ultra and related experiments to create a programmable subject (though the authors themselves see this not as a turning but as a modification within biopower). These projects vacillate between the attempt to replace one personality with another and the attempt to produce action independent of the subject (e.g., an action contrary to one's deeply held moral beliefs), (Stoneman and Reeves 2025, 111). While the rudiments of control are at work here, the project is broadly one of discipline, of the biopower of individuals in that the primary goal was a programmable subject. The aim was to erase one subject and replace it with another. This program attempted to shape subjects through (usually dubious) procedures such as "de-patterning" to erase the current subject, operant conditioning to implant a new personality, or trying to produce a truth serum to make any subject into a truth-teller, at least temporarily. Importantly, while it aimed at producing new subjects (the dream of discipline), it only succeeded in destroying them (the byproduct of control) without achieving the desired behavioral outputs. Compare this with more contemporary dubious procedures such as forensic brainwave analysis. Here the attempt has nothing to do with subjects whatsoever, but with reading brainwave patterns. Objects are shown to a suspect, some related to the crime, some not. If a "recognition pattern" appears in their brain activity when shown the murder weapon, it is taken as evidence that they are guilty. One's brain functioning is surveilled with no attention to the type of subject they are or could become. A recognition pattern is supposed to be universal across the devious and the honest; it is used not to make someone into an honest subject, but to get at the truth without passing through the subject at all. The point here does not depend on the efficacy or lack thereof of these techniques. Rather, the concern is the way power is attempting to operate. The character, habits, personality traits, etc. of the person are ignored; the concerns of discipline are nowhere to be found. Rather, it is the attempt to record and classify events in the brain, a condition of the possibility of any subject at all. And as we saw with surveillance capitalism's hunt for ever new data, recording and classification are the precursors to prediction, intervention, and modification.

In short then, control differs from discipline in that it operates by externalization, it deskills and it disrupts processes of subject formation, surveillance becomes data gathering and intervention detection (rather than compelling internalization), this surveillance attempts to hide rather than make itself known, and the data gathered is used to model people not as subjects of certain types with capacities, but as dividuals that are constellations whose nodes are potential transformations. Let us now turn to the population level to see how these and other differences also deviate from biopolitics.

3.4 Biopolitics

Biopolitics is "the attempt, starting from the eighteenth century, to rationalize the problem posed to governmental practice by phenomena characteristic of a set of living beings forming a population: health, hygiene, birthrate, life expectancy, race..." (Foucault 2010, 317). It "brings together the mass effects characteristic of a population, which tries to control the series of random events that can occur in a living mass, a [strategy of power] which tries to predict the probability of these events (by modifying it if necessary), or at least to compensate for their effects," (Foucault 2003, 249). Biopolitics becomes an ascendent form of a power when states begin to think not in terms of subjects or citizens, but in terms of biological features, especially in terms of rates (birthrate, deathrate, etc.). Techniques are developed to categorize, manage, and intervene on these features, and such techniques spread beyond the state. Foucault notes that the limit case of biopolitics is Nazism, both through the Holocaust and the self-destruction of German living standards, yet every modern state employs biopolitics to some degree. On this, he gives the cryptic remark "In all modern States, in all capitalist States? Perhaps not, but in capitalist states as much as in socialist states," (Foucault 2003, 260–1).

In short, biopolitics is the half of biopower that deals with populations rather than individuals. In doing so, it has different techniques, aims, and constraints. Foucault's historical overview traces the development of biopolitics as a mutation from discipline to security, or as the change that the forces of discipline must undergo when they take the population rather than the individual as their target. Discipline works in an empty, artificial space that is to be completely constructed (the prison, the hospital, etc.), whereas security, first, works on an already given space (water ways through a city, air flow, etc.). Second, its target will not be reconstructed to a point of perfection; it will be maximized according to the positive elements already there (not shaping the good factory worker but increasing the efficiency of the labor pool). Third, one must reckon with the polyfunctionality of the features of the space (a prison has a single function, a town does not). Fourth, security means accepting an open future that is not precisely measurable, predictable, or controllable, (Foucault 2009, 19–20). Thus optimization here amounts to a mutation of normation into normalization, the latter made possible through statistics (alongside various bureaucratic technologies). While mutating out of or budding off from discipline, biopolitics does not so much replace discipline as become its large-scale complement. This complementary arrangement is biopower, and it is to Foucault's great credit that it is a theory of complex interaction between scales rather than a fallacious "as above, so below" account.

Biopower is ascendent when the state takes on the concerns and techniques of biopolitics. However, that cannot be the whole story because that would be a merely "macro-physical" account, wherein power is reduced to what the state does. This adoption by the state marks the rise of the epoch or the cohesion of a strata, but it does not name all manifestations of biopower. While discipline is generally the small scale and biopolitics the large scale, these diagrams can be transposed across scale. One individual may act on another's conduct in biopolitical ways. Allowing or denying access to a social event based on vaccine status during a pandemic is indeed an example of biopolitics, though unlike certain Foucauldians and conspiracy theorists, the conclusion here should not be that therefore such policies ought be resisted, but that not all expressions of biopolitics are bad. This is analogous to what we have said about control: its techniques are mainly forged by corporations under the model of surveillance capitalism, but they are in the process of being transposed across society at all scales.

Gamez 2023 argues that surveillance capitalism and the various developments in techniques, aims, and concerns of power do not amount to a new diagram, but rather a "particularly worrisome intensification of biopolitics," (87). His defense is largely in response to Colin Koopman's diagnosis of the rise of info-power (which we return to below), though as Koopman's is by far the most worked out of a bevy of other attempted diagnoses at new forms of powers, Gamez's successful defense against Koopman translates to a defense against these other positions as well.

Like we saw with discipline in the previous section, it is not the case that these analyses of biopolitics in the age of big data are wrong. Indeed, the mistake is an invalid inference, from "a known dominant strategy of power has adapted to new materials" to "therefore, no new form of power is cohering." This inference fails for

Foucauldian reasons: strategies of power are not monolithic, all-encompassing, or even mutually exclusive across societies or within institutions. If epochs or strata of power are interpenetrating, overlapping, vague, and are not replaced so much as linger on slowly becoming irrelevant, then we should expect further mutations even as a once dominant form falters. That there are even large-scale innovations in the expression of biopower or sovereign power does not mean that something new has not cohered. Because of this, the question is then not whether the techniques of surveillance capitalism can or are being used in biopolitical ways. For this reason, I won't analyze in detail various analyses of those usages. Instead, the question is whether something distinct has also taken shape.

In the previous section, it was argued that control is marked by the dividual and the mass of data replacing the individual and the population, and that we see this already at work in surveillance capitalism. The dividual and the individual seem similar, in that they are both ways of 'carving' up people, though this is a somewhat misleading heuristic. Populations and masses of data, on the other hand, are obviously radically different sorts of things, the former being collections of people understood as their various biological features (birthrates, etc.), the latter being enormous collections of information of every sort. This difference is best understood not as ontological claims, but rather as a result of the various mathematical techniques that give rise to them. Foucault argues that the rise of biopolitics is inseparable from the creation and adoption of statistics by the state as the science of its own self-understanding (hence the etymology of "statistics"). These techniques make possible understanding (that is, tracking, measuring, and predicting) groups of people in terms of their various rates of change: birthrate, suicide rate, disease spread, consumption per capita, etc. In doing so, it both facilitates a change in focus in what it is to administer a state and it changes the function of norms as well. On the latter, normalization comes to replace normation.

Normation is the external imposition of a standard; normalization allows a standard to rise out of population itself, or rather out of a data set. This process then gives norms a new power; one can use them to move between purely descriptive claims (e.g., "this is the statistical average of the population") and, well, normative claims (where the average sets the standard of what one ought to do). Hence how someone dying before reaching the average life-span of their demographic categories takes on an air of moral failure. Mader 2012 calls this a "sleight of reason" and gives an analysis of it, its errors, and the tricks it plays on us. Again, we should stress this is not the sole province of states; Taylorist management developed much of its techniques via the statistical analysis of workers to produce a (usually overestimated) model of the maximally efficient worker, and the subsequent goal of normalization of factories to shape a population to that norm (see the ambient feeling of never being productive enough).

All of this is still at work. Statistical methods are the backbone of machine learning and fundamental to the development of surveillance capitalism (see Joque 2022 for an account of this through the rise of Bayesian statistics and machine learning). However, this is the earlier strata on top of which new sedimentation is taking shape. Like how statistics makes biopolitics possible, control is made possible by the introduction of linear algebra. To oversimplify, while statistics is the mathematics of collecting, organizing, and analyzing data, linear algebra is the mathematics of transformations. Machine learning is, generally speaking, the result of bringing these techniques together in the realm of programming. What linear algebra makes possible is the encoding of things not as individual data points (like rates on a spreadsheet), but as vectors in a vector space, matrices, and tensors, and then the performance of operations on those objects.

This augmentation in mathematical techniques results in a form of power with a different ontology and a different logic. The ontology of biopolitical statistics understands what one is intervening on as individual particles in Brownian motion (statistical mechanics) or a population characterized by normalized collective rates of change. For control, they are understood as vectors of potential transformation, that is, dividuals. Yes, you may be tracked through surveillance capitalism by a unique identifier that singles you out separated from everyone else, but that is not important, or rather, that is not the innovation. The innovation is that "you" in these systems are not an individual, but rather a collection of vectors of possible transformations, and parts of "you" make up others' vectors. Such a vector is from one state to another, one position to another. A particular vector and the space it is embedded in is not produced from your data alone, but from everyone's data. It is the result of machine learning synthesizing a vast quantity of similar data. A collection of vectors may be tagged with your name, but this is misleading. It is a model of you not as an individual, a finished product, but a model of you in your transindividuality. It models not who you are, but what you may do and what you may become given what forces shape you. Adapting to being treated in such a way is a central force behind the exhaustion of subjectivity. And is it any surprise that we feel ever more pulled in all directions the more our lives are modulated through systems that treat us merely as a bundle of vectors, where following any one vector too far may tear that bundle apart?

Finally, we should note here that there is no reason the class of dividuals would include only individual humans modeled as bundles of vectors. I've focused on the human dimension, and as a form of power, this focus is not out of place, but in general this is a flat ontology. Words, people, countries, protein folds, spreading ideologies, etc. can all be effectively modeled in the same way. Biopower as the micropower of discipline and the macropower of biopolitics follows an analogy with a common view of the sciences, where what is studied is determined by scale, like the relationship between physics and biology, or more directly, psychology and sociology. Control instead is like the approach of complexity science where the same sorts of processes and mechanisms are scale-free, or found at all scales, and are in some way independent of the substrate they are applied to.

As for the different logic, it is important that control's techniques make use of enormous multi-dimensional spaces. The classical use of statistics always comes with a feeling of coarse-graining, an implicit sense that what one is working with is a summary of more complex information. A spreadsheet never feels fully complete, one's calculative predictions always come with caveats about outliers, and so on. This is rarely a check on hubris of course, but it does resonate with the openness of spaces of security that Foucault notes. However, spend time with an n-dimensional vector space that one can explore because the results it generates are not explicitly programmed in ahead of time, and you will begin to understand the casual megalomania of tech bros. They are not justified in their megalomania of course, as seen when any humanities undergraduate reminds them that the map is not the territory, but given that one is working with a seemingly infinite map containing shockingly powerful predictive structure, the pathology is understandable if not excusable. The fantasy at the heart of this pathology is that the perfect map is not just possible, but achievable if only we had a bit more data. If so, then the future is not open, but neither is it deterministic; the future is shapable via prediction and intervention. Note the ever-shrinking margins of error, for example, of an Amazon factory worker who is tracked, prompted, and evaluated down to the second so as to reduce task times to their absolute minimum, while designing related systems to operate assuming this minimum is continually met. Whereas biopolitics, with its focus on security, accepts an open future, surveillance capitalism does not. This is a major portion of Zuboff's analysis and why she champions a "right to the future tense" as a form of curtailing surveillance capitalism. No such right makes sense as opposed to biopolitics.

This also clarifies why norms do not have a central role in this new form of power, at least not in the fundamental way they do in biopower. Descriptively, a norm is always a coarse-graining, something we use in place of more detailed information because that information is unwieldy. A norm is a weight on behavior, a way to sanction or permit through social pressure, often when there is no other way to check behavior or to avoid the expedience of violence. We see both at work in how out of step norms often are with actual human action, motivation, and flourishing. Yet, if one has access to the complex products of machine learning, one does not need coarse-graining; the algorithm wields the fine-grained information perfectly well, regardless of accuracy. Nor does one need to worry about others learning and internalizing norms. The actions of others are presumed shapeable in a predictable way through targeted intervention.

Finally, we should also consider that the central motivation of biopower, to maximize the potential of life, is falling ever more from the concerns of the state. Our age is marked by a dereliction of response to climate change, a threat to human life itself. We are in the wake of a disastrous handling of a pandemic where the American state's main response has been to continue to gut its own biopolitical capacities. Foucault constantly returns to health and hygiene as main concerns for biopolitics. The covid pandemic made it clear that whatever lingering biopolitical forces there are (and insofar as there are populations and humans have biological conditions that intersect with governmental concern, they will linger), it is not the dominant form. There was instead a question of how many deaths could be borne? What is the loss of life acceptable to maintain the economy? Yes, talk of culling the weak is biopolitical through and through, and we heard plenty of such discourse throughout, but this is not what state capacity mobilized for, nor was it even a particularly dominant form of discourse. Instead, state capacity was wielded to oppose the gain in power by workers and to try to hold together a global supply chain and economic system that, in the name of efficiency and with a faith in prediction, spent the last several decades excising its margins of error.

3.5 Infopower

Gamez notes Koopman's analysis of infopower as the most worked out attempt to diagnose these technical, economic, and political changes as a new form of power beyond biopower. I take this to be broadly correct. Gamez's argument is that infopower does not meet the standard required to be considered a new form of power, and since the best candidate of the new powers camp is rejected, we have reason to conclude that what we are seeing is still biopower. Having made the case that biopower does not explain these new mutations and that we are instead seeing a new diagram of power being transposed, do I then need to work through all the other various candidates to show that control is something different? This would be redundant, because both Gamez and Koopman give compelling arguments against treating any of these as characterizing a complete new form of power, (see Koopman 2019, 161–171). However, it is necessary to engage infopower here for two reasons. First, Koopman gives an account of a mechanism central to understanding this new form of power and its historical roots: fastening. Second, his argument against control should be addressed.

Fastening is Koopman's term for the process by which we become data-ified. Data is not simply available and waiting for us to collect it; raw data is a contradiction in terms. To exist, data requires there to be classificatory systems, systems of collection, of retention, and so on. Fastening is then the processes that make regular data collection from humans possible. Here Koopman makes use of both senses of "fast" in fastening; to make a strong connection (to "hold fast" or be steadfast) and "fast" as in speed or to speed up. Our data, and the various systems that make that data possible, stick to us, become fastened to us, and in doing so, speed up the ways in which we can be recorded, processed, and modulated through various systems. While Koopman might not put it in these terms, fastening is the process of making the traces of human behavior ever more legible to and retainable by technical memory. This is an important insight into the functioning of this new form of power, and much of Koopman 2019 is a genealogy of this process through various historical antecedents.

This historical methodology is at the heart of Koopman's critique of control. Deleuze and the authors who pick up his work, Koopman says, are engaged in a mistaken avant-gardism. Control, he continues, is focused on a radical break with the past, and neglects the historical work necessary for understanding this form of power, (Koopman 2019, 169–171).[5] There is here a correct point entangled with a misunderstanding. First, the misunderstanding: to focus on this new form of power

[5] Koopman also notes the paranoid quality of talk of "control," which I address in the Introduction.

as new is not to claim it is a radical break without historical precedent or that historical study is thereby unnecessary. While the followers of Deleuze may focus on the new and the unknown of the future, nothing of the account hinges on this focus. The newness is not a fact of all the components, but of the diagram and its ascendency. A discrete shifting of regimes and styles of power are grounded on a continuous shifting of what is possible and what has been done. I agree with Koopman's claims about the need for history and the rigor of genealogy here. Such a genealogy would be as massive as it is necessary, including much of what we addressed only in brief in the previous chapter alongside not yet fully understood connections. Such further connections include those that are currently ongoing such as the various boomerang effects of techniques developed for anti-insurgency,[6] the development of blockchain and zero trust infrastructure in cybersecurity influencing governmental system design, the use of generative AI to dismantle the workforce, and so on.

However, to find earlier instances of certain techniques is not to establish that a diagram of power exists then. The clearest case of this is in Foucault's own analysis of biopolitics, a story which for him begins in the Middle Ages with Clovis (Foucault 2003), yet biopolitics on his account is not formed or ascendent until the eighteenth century. I take it that the genealogy of techniques that Koopman expertly undertakes has this same relation to control. The supposed avant-gardism thereof is not inherent to the position but to the fact that Deleuze is calling for a new project, one that requires, among other things, both the production of historical understanding and grounded speculation. Yet we should be careful not to overvalue the historical, lest one always end up fighting the previous war. Nothing has a history until it is born (and often not even then).

This undercuts Koopman's argument against control, but in a way that makes his project an ally. This has been my approach to the other thinkers of new forms of power whose insights I have made use of, and it is an approach that if successful, means further arguments over whether we should call this "control," "infopower," "datapower," "neuropower," etc. are merely semantic in the pejorative sense.

3.6 Control

We have one task left to us in this chapter: to survey the diagram of control. We must not only describe the nature of control; we must also clarify limits, failures, and even features of control useful for positive projects, else it would fail to live up to the standard so far set for an account of power.

Control is a diagram of power that uses masses of data to map vectors of dividuals via continually modulating relations of access. The dependence on masses of

[6] See, for example, Chap. 7 of Trinquier 1985 which calls for the creation of a total information network to fight insurgent groups. This comes along with an admonishment that any country that does not do so is negligent regarding the threat of invasion; "there is no excuse," (34–5). See also the Total Information Awareness project as a response to 9/11.

data pushes towards ever more expanding and centralizing of technical memory, and it is through information technologies that this expansion becomes increasingly more complex and automated. This automation happens to the point that digital artifacts may autonomously anticipate and respond to expectations, which we make a part of our goings about the world and our cognition. That is, it gives rise to what Hui 2016 calls tertiary protentions: certain artifacts anticipate and fulfill certain responses without direct human prompting. This makes collecting ever more data an imperative (in the large scale of corporate data extraction as much as the interpersonal scale of taking hundreds of photos one never looks at again or screenshotting conversations, of "having the receipts" just in case). Such collection is not just collecting more of the data we already know how to use, but also to continually create new forms of fastening, that is, new ways of data-ifying people and things. This imperative gives increasing recording, surveillance, transparency, and access a quasi-moral status. All the worst for subjectivity.

Modulation means here a constant change of available options, possibilities, opportunities, and the like. This is what allows control to operate without the enclosures of disciple or the constant re-normalization of biopolitics: the open spaces are mapped and structured ahead of time and continually remapped in real time. Deleuze's example of the highway as a pre-planned structure of movement that nevertheless gives the feeling of freedom morphs into the information superhighway. What is being modulated is not just access to resources, spaces (physical and digital), but also access to and via technical memory itself. Insofar as human cognition depends on technical memory, this gives some degree of power over what cognition can access, can make use of in its functioning. This operates not only at the level of conscious thought but also at the level of psychic and social individuation and does so in ways that disrupt both.

It is worth reiterating that to say we move within spaces that are suffused with power is not to say that these forces are all-powerful. This is to conflate power with coercion. A highway does not in itself determine my destination; it does however weight each choice of destination by the degree of access it provides or denies. This is in part why the central danger of this diagram of power is the exhaustion of subjectivity, the human with a massively damaged or destroyed inner life, not the programmable human being perfectly guided by microtargeted interventions (even if the latter remains a guiding fantasy as well). Furthermore, a diagram is never fully implemented and is never total. Nor are its prescriptions always effective. Nudges are in many ways a paradigm tool of control in that they are modulations of ease of access that preserve the feeling of choice. They are, however, often completely ineffective.[7]

---

[7] See for example Arshad et al. 2019 which shows that there is effectively no difference between countries that design their organ donation forms to be opt out rather than opt in. Nudging people to become organ donors by requiring extra effort to opt out and making opting in the easier task is often touted and accepted as an "obvious" case where nudges produce clear benefit. No such obvious fact is borne out by the data. This is not to say that all nudges are ineffective, and we should

Finally, if one accepts that a form of power is here to stay, then it is a mistake to focus solely on its harms. To paraphrase Deleuze, we need neither fear nor hope, but only to seek new weapons. Some inspiration is to be found in Latour's admonishment to Foucault. When speaking about social alliances and changes in response to the germ theory of disease, he notes those who "use the microbe to redefine the celebrated 'self-interest' and to link everybody together through fear of disease. This unexpected strengthening is not in itself 'reactionary,' as suggested by some authors who are used to speaking only of power and who see hygiene as a 'means of social control.' The allies of the microbe are to be found on the left as well as on the right," (Latour 1993, 36). Judging whether this is fair to Foucault or better said of the more Rousseau-inflected of his disciples, I leave as an exercise to the reader.

The point remains: if we are going to be able to administer vaccines to populations, especially to the percentage needed to achieve herd immunity and protect those who cannot take them, then we are going to accept a certain level of biopower. We ought to. "Defeating" biopower is a silly idea; an anarcho-primitivist fantasy to undercut the very technological conditions of current populations, a fantasy whose realization would amount to a death toll that dwarfs any twentieth-Century genocide. If two common Foucauldian notions are true, 1: power is not just repressive but productive as well, and 2: these forms of power are neither bad nor good, but dangerous, then the focus on a mostly formal imperative to resist is deeply misplaced. Rather, understanding the diagrams of power at work in one's world, which are being formed and transposed, which are breaking down, and so on, is understanding the background conditions in which one moves, acts, and is shaped. This realization, of course, comes with its own dangers. It risks capitulation, but we will return to this problem in Chap. 7.

Here we should survey briefly the opportunities and the new weapons of control. I exclude here the advantages of specific technologies or of the information age generally, of which there are too many to count; rather, I point to those inherent in the diagram of control itself. First among them is that the retreat from shaping interiority opens some possibilities of creativity. This happens alongside the weakening of many of the harsh taboos undergirded by the norms of discipline. Experimentation is freed to a wide degree. Furthermore, even in its central danger, the exhaustion of subjectivity, is a potential boon, if it can be limited. The self is not all it is cracked up to be. Much personal distress is found in realizing that one is not the type of person they want to be, that the subject they have been shaped into has habits one finds harmful, upsetting, or immoral. Insofar as this form of power both breaks down subjectivity itself and makes new ways of intervening in one's own ongoing processes of subjectivation, radical new possibilities of self-transformation emerge.

Interestingly, it is here that we see big tech most limit the liberatory powers of control. Under discipline, a class of workers must be produced that is competent, capable of working together, etc. in order to complete factory work. These are also

probably treat, say, paper form design differently from micro-targeted, big data and surveillance-enabled interventions.

capacities that make effective unionizers, and so mitigations are produced and structures to channel these capacities are enforced. In the same way, control institutes just enough skill, subject formation, and bulwarks for self-cohesion for one to be manageable. Take, for example, the real name policy of some social media like Facebook, compared to the wild experimentation and self-shaping possible on the earlier internet. Those of us who experienced the internet before it was colonized by a handful of social media platforms and Amazon understand the ecstasy and vertigo of the *Cyborg Manifesto*'s embrace of cyberspace to challenge identity; those raised on a platform-controlled internet often do not. Many have claimed that de-anonymizing the internet would lead people to act more responsibly. I think we have good evidence that the opposite is frequently the case: by tying every ephemeral online experimentation publicly and permanently to one's "real identity," an immense pressure to double-down on one's worst expressions was created.

Graeber 2016 raises a point that has an interesting implication for control. He notes that violence is one of the most effective ways for humans to have predictable effects on others without knowing much or anything about those others. If I want someone to stop what they are doing, negotiation requires I know something about their wants, motivations, their capacities, etc. Shooting them, on the other hand, requires knowing nothing beyond their physical location. Control is the attempt to have predictable effects on others through knowing as much about them as possible. Power is always grounded in the possibility of violence, and control makes terrifying new forms of violence possible. However, in the imperative to gather greater information to influence behavior, control also has at least a structural tendency away from the direct use of violence.

We should not be surprised by these potential benefits. Control is a problem for Deleuze especially because many of its tactics are what he and Guattari in the two volumes of *Capitalism and Schizophrenia* (and in different ways many other great thinkers of the twentieth-Century) took to be inherently liberatory. Modulation, horizontal organization, casting off the self, transgressive breaking of taboos, desire freed from discipline, etc. Each of these is either a result or a tactic of control. In particular, the subject, unmasked from its disguise as a sovereign atomic individual or a soul, was understood to itself be a sort of technology of managing people. *Anti-Oedipus* directly aligns undermining the conditions of fascism with throwing off the tyranny of the self. The dangers of control show taking these tactics as unqualified goods to be a grave theoretical mistake. It does not, however, show the critiques they followed from to be false. What is then needed is analysis of what about interiority is worth saving, which is the work of the next chapter.

References

Arshad, Adam, Benjamin Anderson, and Adnan Sharif. 2019. Comparison of Organ Donation and Transplantation Rates between Opt-out and Opt-in Systems. *Kidney International* 95 (6): 1453–1460. https://doi.org/10.1016/j.kint.2019.01.036.

Deleuze, Gilles. 1997. *Negotiations 1972–1990*. Translated by Martin Joughin. Columbia University Press.

de Laat, Paul B. 2019. The Disciplinary Power of Predictive Algorithms: A Foucauldian Perspective. *Ethics and Information Technology* 21 (4): 319–329. https://doi.org/10.1007/s10676-019-09509-y.

Foucault, Michel. 1990. *The History of Sexuality, Vol. 1: An Introduction*. Reissue ed. Vintage.

Foucault, Michel. 1995. *Discipline and Punish: The Birth of the Prison*. Translated by Alan Sheridan. Vintage Books.

Foucault, Michel. 2003. *Society Must Be Defended*. 1st ed. Picador.

Foucault, Michel. 2009. *Security, Territory, Population: Lectures at the Collège de France 1977–1978*. 1st ed. Translated by Graham Burchell. Picador.

Foucault, Michel. 2010. *The Birth of Biopolitics: Lectures at the Collège de France, 1978–1979*. 1st ed. Translated by Graham Burchell. Picador.

Gamez, Patrick. 2023. Inhuman Hermeneutics of the Self: Biopolitics in the Age of Big Data. *Foucault Studies* 34:80–110. https://doi.org/10.22439/fs.i34.6939.

Graeber, David. 2016. *The Utopia of Rules: On Technology, Stupidity, and the Secret Joys of Bureaucracy*. Melville House.

Han, Byun-Chul. 2017. *Psychopolitics: Neoliberalism and New Technologies of Power.* Verso.

Hui, Yuk. 2016. *On the Existence of Digital Objects*. Univ Of Minnesota Press.

Joque, Justin. 2022. *Revolutionary Mathematics: Artificial Intelligence, Statistics and the Logic of Capitalism*. Verso.

Koopman, Colin. 2019. *How We Became Our Data: A Genealogy of the Informational Person*. 1st ed. University of Chicago Press.

Latour, Bruno. 1993. *The Pasteurization of France*. Translated by Alan Sheridan and John Law. Harvard University Press.

Overwijk, Jan. 2025. *Cybernetic Capitalism: A Critical Theory of the Incommunicable*. Fordham University Press.

Patton, Paul. 2018. Philosophy and Control. In *Control Culture: Foucault and Deleuze after Discipline*, ed. Frida Beckman. Edinburgh University Press. https://www.jstor.org/stable/10.3366/j.ctv7n08g6.16.

Reeves, Joshua, and Ethan Stoneman. 2025. Brain Warfare and the Malleable Mind: Experiments in the Programmable Subject. *Theory & Event* 28 (1): 97–119.

Seecamp, André, and Jan Söffner. 2024. A Postscript to the Postscript on the Societies of Control. *Substance* 53 (2): 75–85. https://doi.org/10.1353/sub.2024.a934263.

Trinquier, Roger. 1985. *Modern Warfare: A French View of Counter Insurgency*. Translated by Daniel Lee. Pall Mall Press.

Chapter 4
Interiority Beyond the Subject

> *"Between the impulse and the act, the tiny interval that is reflection.*
> *Where there is no room for reflection, there is no room for justice or prudence."*
>
> *—Simone Weil, The Iliad, or the Poem of Force*

Post-structuralist thought is either ambivalent to the self or outright hostile to it. The breakdown of the domination of the subject, the individual, and the greater enmeshment into processes of individuation that control undertakes is in a perverse way what Foucault, Deleuze, and many other post-structuralists called for. One might well wonder if taking these figures seriously then means embracing control fully and welcoming the exhaustion of subjectivity. Does it mean that these figures' right wing accelerationist interpreters are their true heirs? I think not, though these thinkers are also not capable of fully responding to the challenge posed by control alone. This does not mean their critiques of the subject are wholly mistaken. Rather, where the subject is understood to maintain identity, recognition, current power structures, and the illusion of the isolated, autonomous individual, these critiques are broadly correct. However, the exhaustion of subjectivity does not threaten only these dimensions of the self (and in some ways intensifies them).

Rather than autonomy, I will argue that much of what is harmed by control is our capacity for heteronomy, that is, for the possibility of productive encounters with that which is outside ourselves. Subjectivity as interiority is a condition of heteronomy as an individual must be constituted for there to be an inside that can encounter something beyond itself (recall that individuation involves the production of a distinction between inside and outside). This distinguishes heteronomy from transindividuality, a distinction rarely made by thinkers of individuation. Stiegler, for example, conflates these levels often in that he treats all communication, especially through text, as happening at the level of social individuation, (see for example, Stiegler 2010, 24–5). Excluding either or reducing one to the other will only lead to

M. J. Ardoline, *Interiority Against Control*, Sophia Studies in Cross-cultural Philosophy of Traditions and Cultures 47,
https://doi.org/10.1007/978-3-032-16245-8_4

flawed views of human flourishing.[1] This chapter will then argue that any harm to interiority as such is a harm to heteronomy and what it makes possible, including an appreciation for truth, the organization of desire, and ethics in general.

4.1 Critiques of the Subject

The struggle of late twentieth-Century continental thought was in many ways the struggle against the self. The self or the subject was seen as a sort of straight-jacket or a thing invented to simultaneously be policed and do some of the policing. The most influential version of this is perhaps Althusser's account of interpellation, where we go about our day lacking anything like a grasp of ourselves as a subject. We only gain one when the cop shouts "hey you!" and we recognize or misrecognize ourselves as the one called. The subject then is literally a part of the police apparatus. Because of this, "Poststructuralism has tended to romanticize the loss of self," (Howells and Moore 2013, 10). Take for example this declaration from Foucault: "I am no doubt not the only one who writes in order to have no face. Do not ask who I am and do not ask me to remain the same: leave it to our bureaucrats and our police to see that our papers are in order," (Foucault 2012, 17). Similarly, Deleuze and Guattari counsel us that whatever a cure may be for the suffering of the schizophrenic, it will not be the restoration of the self. Rather, the "I" is one delusion from which the schizophrenic is free (Deleuze and Guattari 1983, 23). It is hard not to sympathize.

These critiques have often been taken to imply that a loss of subjectivity both frees one politically and undermines the condition of individual suffering. These readings equate complete disorganization with freedom and a lack of being identified with being immune to political oppression (as if a gun could not fire without knowing what it was pointed at or the state could not categorize one anyways). The exhaustion of subjectivity puts the lie to this. Even if one is psychologically disorganized, they can suffer; even if one has abandoned all identification, as long as data can be fastened to them, they can be targeted. To the degree that these readings are accurate (and that is not an insignificant degree), control is a problem that poststructuralism lacks the resources to respond to adequately. Control has many features of the positive possibilities these thinkers saw politically (decentralization, horizontal organization, breaking of norms, etc.) as well as the undermining of the subject. Deleuze and Guattari clarify and modify this in the shift between *Anti-Oedipus* and *A Thousand Plateaus*. However, despite the admission that not all trees (hierarchies) are bad, there are good uses of strata, and an embrace of self-cultivation, even in the

[1] This distinction is analogous to that made between intersubjectivity and ethics in Métais and Villalobos 2022. The difference between the two is that they are opposing Levinas to other phenomenological ontologies and oppose a version of intersubjectivity that maintains the individual, whereas I am grafting the same point onto Deleuze's metaphysics and so emphasize that heteronomy cannot be found in continuous process alone.

later work there remains a confidence in the liberatory power of the always mobile, the nomad, the rhizome, that state capture can be avoided through speed, and a self will only slow you down. In describing control societies, Deleuze seems to realize this confidence was mistaken. The "Postscript on Societies of Control" poignantly hands the problem of diagnosing and responding to control to future generations. This is at least as much an admittance that he does not have an adequate response to it as it is a humility that a form of power must be lived through to be understood, and as a recognition of his own impending mortality. This does not mean the critique of the subject is mistaken. Far from it. That control does what it does shows the account of subject formation and its possible disruption is correct. However, it also shows that the liberatory potential of that disruption is not straightforward.

The focus of the poststructuralist critiques of the self is then how subjectivity 1) is and is used to maintain a stable identity that makes one legible to powerful institutions, and 2) produces in oneself the illusion of being completely separatable from or transcendent to the world around them such that they are an atomic, autonomous individual. As simple as these seem, they have long histories and vast consequences politically and psychologically. If one is separate, autonomous, and stable over time, then they can be made responsible. The making of a subject intersects with the machinic phylum as well as with techniques from punitive scarification, branding, and tattooing to criminal records and GPS house arrest tracking, all of which work to stabilize a recognizable, responsible subject. Control accelerates all of this by 1) massively improving tracking and recording of individuals, 2) increasing individual capacities (e.g., accessing much of the world's knowledge instantly or summoning others to do one's bidding via Uber Eats) nurturing a false sense of autonomy, 3) increasing atomization and thereby the feeling of separateness, and 4) ever increasing, usually public, recordings of one's actions and deeds alongside awareness of world events, and so astronomically ratcheting up what one can be made responsible for. If this were all interiority was, then it would be worth discarding along with the soul.[2]

I won't comment on Foucault or Althusser, but I take the reading of Deleuze and Guattari as radically opposed to subjectivity or the self to be a misunderstanding, though one they admittedly exacerbate. Were it accurate, they would not provide a guide in *A Thousand Plateaus* for building a BwO, but rather one for dismantling it. Whence the misunderstanding? It follows from conflating three things: a common, classical notion of the subject in the European tradition, the way power operates on people through processes of subjectivation, and psychic individuation as such. The first two are allied in that the notion of the atomic, autonomous individual is often used to justify and obscure the effects of subjectivations that make one legible and vulnerable to states, institutions, etc. Understanding individuation properly should oppose one to both. The psychic individual is the result of a process that is mostly not in its own power and that it ontologically depends on. It is then not transcendent

[2] Note also how the soul is one more example of Stieglerian spiritualization of technology: the soul is taken as the perfect recording surface of one's deeds and the storehouse of one's true natures, tasks actually achieved by technical memory.

to the world. One's autonomy can only be a result of this, not a fundamental feature of a person's nature. This process is intertwined with social individuation, so one is not isolated but is always to some degree suffused in the transindividual. One's various identities and the capacity to recognize them are not inherent, unchanging features, but temporary, metastable formations that are scaffolded by technical memory. This is the proper understanding of the critique of the subject.

This proper understanding is dependent on the account of individuation. As we have said several times, individuation is a process with a product. That process changes our understanding of the nature of the product, but it does not explain away the product as such. This is the error of the "process only" readings of Deleuze. These readings reject the formation of individuals, yet they embrace novelty as a source of liberation and the good. These are flatly incompatible claims. Novelty in Deleuze's metaphysics rests on the productive power of interaction across and through difference itself. Yet if everything were merely continuous processes flowing into each other, there would be no abysses of difference to cross, and so no power to produce novelty. Furthermore, something new is produced only when it becomes discontinuous to some degree with its milieu. The new is always the result of a process of individuation. To put a stop to one's various individuations would be either death or complete stasis, as the conditions of something new being produced would be undermined. The critique of the subject then cannot amount to the rejection of generating new psychological and social individuals. This is clear throughout *A Thousand Plateaus*, including the claims that there are worse things than remained striated, absolute deterritorialization is death, and an entire plateau dedicated to (re)building one's own subjectivity.

In working this critique out, we must be careful of any uncritical fleeing back to the subject as a reactionary impulse, one no more useful than technophobia. Rather, we must work through it to isolate what is the real target of these critiques, and what, if anything, is worth saving. Here we must separate what the subject does from the illusion it projects of itself. First, the subject is a separation of inner from outer, in both psychological and, to some mutable degree, biological senses. Second, it is a BwO, that is, it is a part which organizes other parts. In doing so, it provides a metastable structure to them, providing cohesion and making coordination possible. It is also a recording surface, retaining traces of what these various parts do and what happens to them (Deleuze and Guattari 1983, 11). It is these two features combined that create the illusion of the classical autonomous subject. A general is the BwO of an army, one soldier among many who plays the role of organizing. A general can be replaced and the army can continue or disperse. In the battle, the general neither fires every weapon themselves, nor do soldiers follow the general's orders exactly. Yet, battles won and lost are recorded on the general. We often speak of a general winning a battle ("Wellesley won at Waterloo") as if they were able to do so themselves. This is because the organizing function of the general is paired with the inscription of traces, leading to the illusion that it was the general alone that was responsible, as if the organizer had complete command and control of the organized parts. Hence why Deleuze and Guattari refer to the miraculations of the BwO: it would truly be a miracle if Wellesley himself defeated Napoleon's army, yet we find it hard not to discuss or think about the battle without seemingly implying such a miraculous event. Such is the transcendental illusion of the BwO. It is exactly the same, not metaphorically the same, with the subject and the biological, psychological, social, technical, etc. parts it attempts to wrangle.

If we are both process and product, then there are two levels on which to be concerned with the exhaustion of subjectivity. The level of process is that of our various individuations, subjectivations, and transindividuations. We have already surveyed some of what is lost at this level when it is disrupted by control. At the level of product, of ourselves as individuals, what is lost that is most important is not any of the features we see in the critique of the subject. Rather, it is heteronomy and all that follows from it. Insofar, then, as there is a positive use of interiority, organization, and inscription for human flourishing, the subject is not to be cast aside completely. The illusion of the atomic, autonomous subject, however, must be discarded as fast as possible (how much suffering is produced because one thinks they should be able to command everything which they recognize as a part of their psychology, biology, or sociality?). The focus on interiority for the sake of plumbing the depths of the self, from Augustine through Rousseau, the Romantics, and the Existentialists with their authenticity was a mistaken narrowing or misunderstanding the point of interiority. The subject and its relationship to organization and inscription will be the focus of Chap. 5.

Briefly, we can also look at another major tradition critical of the self: Buddhism. Accounting for the various ways different schools within Buddhism interpret the no-self doctrine and its relation to both the exhaustion of subjectivity and information technologies more generally would take an entire work of its own. Here I want to only make two points (for the sake of brevity, I will try to keep the discussion at the level of what many schools of Buddhism take to be conventional rather than ultimate reality). First, no-self is not the exhaustion of subjectivity, and they must be opposed in some important way. This can be established phenomenologically. The practices which facilitate greater recognition of no-self are generally associated with positive attainments and affects, such as greater compassion for others, increased capacity for staying calm, and a reduction in suffering. The exhaustion of subjectivity is quite frequently the exact opposite. It is a feeling of distraction from the self or a loss of self that results in less emotional stability, feelings of guilt for engaging with it, decreased empathy, and so on. Despite what we might think given how compelling it is, no one has yet reached enlightenment through hours spent mindlessly doomscrolling.

Second, even if no-self is not the exhaustion of subjectivity, even if it is opposed to it, we should be skeptical that achieving it would be effective for opposing the exhaustion of subjectivity. No-self follows from the Buddhist doctrine of dependent origination, that all things are dependent on other things, that things are empty of anything like inherent substance, essential nature, or independent existence. It is a form of the process-only view. If this is the case, then (and here we are speaking, as well as one can, of the level of ultimate reality) there is no such thing as interiority. If so, then they cannot account for those things which are lost when there is no interaction between individuals, only those of the interweaving of processes.[3] In short, no-self and dependent origination, like other process-only views, cannot account for the positives of heteronomy, and it is precisely these that are worth safeguarding against exhaustion. Because of this, I am doubtful that any school of Buddhism can both maintain no-self and dependent origination and yet offer a consistent critique of the exhaustion of subjectivity. For the rest of this chapter, we turn to interiority as the condition of heteronomy and how this understanding of the subject recovers and changes several goods classically associated with autonomy in the Western traditions.

4.2 Heteronomy

Individuations result in separations between inside and outside. These separations are porous, metastable, and contingent; they are not transcendent separations. Psychologically, this is the production of an "inner space" that we perceive directly whereas others perceive only indirectly through our actions, expressions, etc.[4] It is exactly this opacity that control qua behavioralist engineering seeks to reduce or eradicate. Nor is this a new desire. It is a deeply human yearning to be more transparent to others ("if only I could show them how I really feel") and to desire others' transparency (hence the perennial fantasy of mind reading and related psychic powers), not to mention to ourselves (hence the impulse to "find oneself"). Rousseau, for example, claims humanity had its golden age when we were transparent beings to each other (Cohen 2008, 88). The more we sympathize, the more we should see this as misguided.

Kant unravels Rousseau's fantasy. He notes two forms of mental opacity: our opacity to ourselves and to others. Opacity at first glance seems entirely negative. It makes possible deception, betrayal, self-misunderstanding, and the like. On Kant's view however, opacity is a condition of sociality, (Cohen 2008). To make his point, Kant imagines beings he calls "sincere aliens" who are completely transparent to others and asks what their societies would be like. He concludes that "Unless they are all as pure as angels, we cannot conceive how they would be able to live at peace with each other, how anyone could have any respect for anyone else, and how they could get along with each other," (Kant 2006, 250, (7: 332)). While not a perfect instantiation of this thought experiment, one could see the discourse resulting from the confessional nature of social media as confirming evidence. Kant's argument then is that insofar as we are not perfectly pure, harmonious beings, the mental

[3] This comment is in the lineage of Neo-Confucian critiques of Buddhism as inherently selfish despite its claims of no-self because there is no real place for people radically distinct from ourselves.

[4] I say perceive here to imply that we are not transparent to ourselves, we do not understand everything in our inner life, and we are open to all the pitfalls, illusions, and errors in inner perception that we are in outer perception, and perhaps more.

opacity of ourselves and others is necessary for society, peaceful companionship, and ultimately human survival itself. One might find this surprising given Kant's ethics requires that we do not tell lies. Here Kant makes a distinction: lying is when we are actively dishonest, but holding to social conventions instead of disclosing all the various facts and feelings of our inner life is pretense.[5] We have no categorical obligation to avoid pretense. It is because we are not already harmonious that we must produce social practices, institutions, and so on. Mental transparency would undermine this.

The defense of opacity and interiority when undertaken, however, is usually paired with, or seen as following from the atomic, autonomous subject. This can be either a metaphysical claim that we really are such subjects, that is, souls, or that autonomy is a fundamental value which must be defended at all costs. This is not only in Kant, but in contemporary technology ethics as well. As new technologies, such as brain-computer interfaces, raise the possibility of piercing mental opacity, the literature on the ethics of mental privacy and integrity has grown. Zuk 2024 provides an overview of this literature and a compelling criticism of its exclusive focus on autonomy. The argument is essentially that there are aspects of our inner lives in which we are not autonomous but which we should value not being interfered with (in particular, our affective states, our epistemic capacities, and certain non-autonomous forms of volition). That is, our concern for mental integrity should extend beyond violations of our autonomy.

In general, I agree with both points above: opacity is a condition of sociality and non-autonomous aspects of our inner life are fundamentally important to us. However, neither goes far enough in answering the question of this chapter: if the critique of the subject (in its proper form given above) holds, then what about the subject is worth saving from its exhaustion under control? For sociality, control makes us an offer: empty out the self, become transparent, and let the organization of society be done by technocratic solutions. The techbro belief in a frictionless society is, whether they would acknowledge it or not, just the belief that we can be made into Kant's sincere aliens. As for mental integrity beyond autonomy, this risks instead intensifying the atomization of the subject in that we may come to understand mental integrity and the defense of interiority as the ever-increasing demand for non-interference with our inner lives. This is a fantasy insofar as minds are not closed containers radically separate from the world.

What is needed to answer our question then is the existence of important goods that are conditioned on opacity, that is, interiority, and on separation but not atomization. Separation is the result of a porous boundary, and opacity means that the crossing of that boundary is only ever partial. By partial, I mean that certain boundary crossings carrying the signs of opacities beyond our own, in excess of our interiority, and which cannot be brought into it without changing our interiority or

[5] In the famous Kantian example of the axe-murderer asking if you spouse is home, presumably interactions with an axe murderer are rare enough to not have attached social mores. It would then be lying rather than pretense to give false information.

destroying that external opacity. This condition is heteronomy, and it is the source of the goods of the subject that are fundamentally worth defending against control.

The twentieth-Century thinker who has most thoroughly worked out the importance of heteronomy is Emmanuel Levinas. There are deep disagreements and theoretical differences between Deleuze and Levinas, more so than between Deleuze and Foucault, Simondon, or Stiegler. Given this project's commitment to a metaphysics of difference as worked out by Deleuze, we are more constrained in what insights we can take from Levinas. The most important of Levinas's insights follow from a simple structure. As humans, we have some partial separation from our world and environment, some interiority that sets us apart. This is the sphere of the Same, or the self, (Levinas 1969, 102). This interiority is different from but nurtured by its environment (Levinas 1969, 111). It is finite but is structured such that it attempts to totalize. Totalization is the attempt to make sense of everything the self encounters in terms of its own projects, goals, concepts, etc. But every so often, it encounters something that cannot be subsumed under these categories. It can only be understood partially, as that which is always beyond the self, as that which cannot be internalized or made a part of one's world without destroying what is unique about it. One can at best be responsive to it, (Levinas 1969, 195). This is because of the interiority of the other, of their excess over whatever I may think of or plan for them. This interiority cannot be experienced directly, but is only ever indicated by a sign, what Levinas calls the face. The face disrupts my projects, my goals, my beliefs, and so on. This disruption, however, is not a mere frustration, but a productive encounter that makes a new form of subjectivity, ethical subjectivity, possible. One could sum this up by saying that for Levinas, the problem of other minds is not a problem; it is the ground of ethics.

What is particularly useful for our project here is that this opposes a fundamental assumption of much of the Western philosophical tradition: the association between knowledge, light, and the good. This is the metaphysics that underlies the compulsion of confession and equates what is with what can be known (and so reduces the existence of the other to our knowable encounters with it; the metaphysics of light is inherently totalizing). Sparrow 2013 explicates this underappreciated portion of Levinas's critique. "The problem with this perspective is that it restricts the realm of sense to what appears and has form; it disallows for meanings or significations that resist the ordering gaze of the ego," (Sparrow 2013, 10). What is lost in this is all that is good which lives in the night, all that can only be encountered as beyond appearance and must be protected from the light to express its existence. "This must not be seen as simply a deficient mode of knowing, however, but as an intimate engagement with a carnal alterity… it is in the night that the caress, voluptuousness, and desire find refuge. It is arguably under the cover of darkness that goodness, as fecundity, is most productive," (Sparrow 2013, 12–3).

I have several times been in a room of social scientists critiquing big data and surveillance capitalism as harshly as anyone. Yet when the discussion turns to what can be learned from such data, they conclude that violations of privacy are justified if they contribute to knowledge production, and lament that if only the data could be shepherded by the trustworthy (read: themselves) but hidden from everyone else, they could learn so much and do so much good. Such an association between light, knowledge, and the good subtends the confessional logic of surveillance capitalism. No critique that maintains this association can succeed. This is not to elevate darkness above light, opacity above knowledge, but to understand that night and day contribute to the good. We continue this argument below where it is shown that without the night of interiority, we may still produce information, but truth loses its meaning. Here we conclude that if it were not for the coming of the other and their excess, if not for the darkness of interiority, heteronomy and all the goods that follow from it would not be possible.

To make use of Levinasian insights that follow, we must first show that the metaphysics of individuation we have been using can accommodate this structure of separation, self, and the coming of the other. This is relatively straightforward as we have rejected the process-only reading of Deleuze. In being individuated, we establish a partial separation from our environment. Separation is the establishment of an inside and an outside (this is the case for both Levinas and Deleuze). As individuals, we are determinate and finite, but as living individuals we carry our individuations with us, never separable from the processes that generate and sustain them. Transindividuation is the processual aspect of ourselves where we are in some sense psychologically, sociologically, and technologically continuous with others (this is the molecular, in Deleuze and Guattari's terms). The way one feels as a dispersed part of a whole during a good concert, the way one's beliefs shift subtly and subconsciously in response to those around them, and the way artifacts extend or curtail our capacities are generally examples of transindividuality. We are never fully separated from the transindividual aspect of our existence, though there are features unique to intense immersion in it, such as collective effervescence, (see Kronsted 2025).

However, individuation fails and ceases if it cannot establish and maintain some temporary, porous separation between inside and outside. In maintaining this boundary, the individual can do things that an indefinite, continuous process alone cannot (e.g., regulate internal temperature). It is also through this porous boundary, here speaking psychically, that others can be encountered as others, as beings radically different than ourselves. What we lose when focusing on individuation to the complete exclusion of the individual is that change can come from the outside as well. There are things that can only be encountered *as* individuals, which change us at that level, and sometimes even reverberate down through our individuations. In mundane encounters, we may still learn from, respond, and adapt to new things. In sublime encounters, we are dis-organized and must re-establish an organization. Both are possible only to the degree that one is first separated from their environment to some degree. These relative changes at both levels and their interaction is what Deleuze calls different/cation, (Deleuze 1994, 209, for a related account in terms of experience in Simondon, see Simondon 2020, 279–81).

Heteronomy, then, is how we are shaped by those around us insofar as we are discontinuous and separated from them. Admittedly, Deleuze is most interested in the sublime sort of heteronomous interaction. He explicitly thematizes those encounters where, in order to respond, we must break down some of our stability and our individuation processes must surge up within us to create something new. In short, he most thoroughly explores those heteronomous interactions that lead to often sudden transformations (see Schreel 2024, Thornton 2025, Ardoline and Cogburn, forthcoming). We will return to this in the next chapter where we maintain such possibilities of transformation, but also explore more gradual forms of self-cultivation as well.

4.3 Ethics

First and foremost, heteronomy is the condition of ethics as such, that is, of reasoning about what is good. On Levinas's account, if we are alone, if the demands of others as living, vulnerable beings do not intrude on us, then we are merely seeking the fulfillment of our projects. This is not bad in and of itself, though it is quite boring and simplistic (not to mention, inadequate to the world given that we are never without others in any true sense). In such a situation, we may not be completely *homo economicus* trying to maximize our pleasure given our beliefs, but our motivations would be quite basic. In the encounter with another, however, I am dealing with something that cannot be fully subsumed into my projects. The opacity of the other means I never have complete knowledge of their desires, needs, goals, etc. They may always be holding something back. Anything that made them completely transparent would mean they were no longer other; they would become a mere extension of the self. Insofar as they remain others, I must consider what they desire, what they need, what they may be hiding, what I know that they don't, and vice versa. This makes possible a capacity Levinas calls substitution (Levinas 1998, 13–4).

Substitution is the capacity to put the other in my place, to reason about what I can do in such a way that I consider others' desires, needs, and goals. This is the production of real alternatives, and it disrupts simple self-interest calculations. The egoist is sure to be skeptical here and will push back by saying that this merely complicates the calculations and only does so to the extent that the other's interests are my interests. The egoist, however, is dependent on the classical account of the subject, and their position fails in two ways. First, the opacity of the other means there is always a deep uncertainty to my understanding of others' interests. Since the other can always surprise me in deep ways, this is not as simple as adding an error bar or uncertainty measure to a calculation; rather, it is incalculable. Second, in our encounters with others, we both have new experiences we cannot have alone and our values and interests change through the encounter (see the section on desire below). This means our encounters with others are always potentially transformative experiences in L. A. Paul's sense (Paul 2016). Paul shows how such cases lead to a breakdown in any current calculative rational decision theory. One could also add that in adjudicating between others, there is often no straightforward calculation of interest one can make. We should note that all this is speaking at the level of individuals; if we take into account social individuation, then any encounter is also the possible production of an "us" which can have values, desires, and interests separate from (or even opposed to) those who make up the us. Substitution extends to this us because one can take up group-interests over self-interests as well.

If Levinas is right, then heteronomy is the condition of a more robust autonomy via substitution. Any account of autonomy that undermines the possibility of heteronomy then undermines itself. The loss of meaningful encounters with others is the loss of autonomy, hence the increase of feelings of passivity alongside loneliness in proportion to that of the false autonomy of atomization. Insofar as substitution changes what I want, that is, changes my motivation to be more than mere self-interest (in fact, it problematizes self-interest itself), it is deeper than Kantian pretense and does not fall into the illusion of the atomic individual, yet it provides a form of autonomy worth valuing (Levinas 1998, 114–5).

That ethics is conditioned on heteronomy and heteronomy requires separation shows what is missing from responses to control that take up the process-only account. The best of these accounts is Striano 2024, which argues for a transindividual ethics. Such an ethics shares much with this project, including the concerns with the damage information technologies can do to subjectivity. However, we differ in that Striano sees the possibility of overcoming this damage through an ethics that functions at the level of the transindividual alone, with the individual being something to be cast off. While I agree that to push against the harms of control, we must focus our efforts on the transindividual level (which is the project of the final three chapters of this book), I disagree for two reasons. First, undermining the subject is central to the harms that control produces. Second, ethics is not possible at the transindividual level alone, for reasons given above.

In closing this section, it is worth noting what this means for recent technological developments, in particular, Artificial Intelligence. AI has autonomy in the sense of being able to produce writing, speaking, action, etc. from an internal state insofar as it acts or operates based on its own internal 'rationality', its algorithmic structure. Whether or not this is connected to any deep interiority is unknowable and immaterial here, even if it is reasonable to believe that for current generation artificial intelligence, especially LLMs, that it does not. What AI does not have in any current architecture is deep heteronomy of the sort described above. If so, then it also does not have the robust sort of autonomy that comes from substitution. What separates human from artificial intelligence, then, is not autonomy but heteronomy. We should not confuse what Levinas means by the face with actual human faces (even if the latter is where one most often encounters the former), and this difference provides a simple example. Facial recognition software is capable of recognizing human faces within some degree of error; it is, however, incapable of encountering the face of the other. The algorithm can run; it can subject images to its totalizing operations of categorization. Perhaps a certain input causes unexpected outputs or the system to crash (such as certain high contrast make-up techniques designed to thwart these systems), but this is a mere glitch or bug, not a disruption in the ethical sense. The system cannot, in an unexpected encounter, suspend the rules it follows and take on the project of trying to understand the good for the other such that it may care for them in ways outside their algorithmic prescriptions. In other words, it cannot achieve substitution. The program can either run or glitch, but it is incapable of opening up to ethical subjectivity. This inability results from a lack of heteronomy. Current generation machine learning is then inherently conservative not so much because it only reproduces the categories of the human worlds it is trained on (this is often true, though not necessarily so), but they are not capable of heteronomous encounters.

I should note that this is not an argument against possible machine consciousness, but rather the claim that any AI with psychic interiority would have this ability (this is likely the sort of ability designing AI along the path of automating workers will oppose as such systems would be less predictable and controllable). Whether looking for this ability could replace the Turing test as a guide for when algorithmic beings deserve ethical consideration is an open question, depending on how easy it is to create simulacra that only appear to do so (on Levinas and AI, see Gunkel 2018 and Gamez 2022).

4.4 Desire

Desire is a fundamental building block of the subject; it precedes the subject and is present in all life at least. Desire as such is not destroyed by the exhaustion of subjectivity, but rather the ability to organize it into a subject is damaged. Here we can make a distinction between molecular desire and molar desire. Molecular desire is all the bits of impulse, all the scattered wants, often un- or subconscious, prior to or outside an organized subject. Molar desire is the result of the organization of the subject, that is, the production of a BwO that can order and record some of the molecular desires one has access to and make them one's own. Like all other organizations, this is temporary and partial. Desire is a particularly anarchic phenomenon, whence both its great potential and the reactionary terror it provokes. Molecular desire is prior to the formation of interiority, and it always threatens stable identities. Desire draws us out of ourselves, making the possibility of new transformations, new organizations, etc. Levinas notes even our metaphysical desire for the other, for that which must always remain outside ourselves, (Levinas 1969, 34). If there is no self to be drawn out of, then we are left with the mere chaos of molecular desire. The same is true if one is drawn out of one's self but is then unable to form new ones.

Were this fact about molecular desire not true, it would be difficult to make sense of advertising. The goal of advertising is either to stoke the desires one already has, or more often, to produce in one a desire they did not already have. The newer the product being sold, the more the goal is the latter, simply because it is unlikely many people wanted something they were unaware existed. This can be done at either the molecular or the molar level; the former is preferred as it more easily bypasses the epistemic defenses of the subject. It is no surprise that advertising has become the backbone of surveillance capitalism as control's ability to intervene at the level of individuation augments this greatly. In this way, advertising is the forbidden art of desiring-production.

The augmentation of desire is worth exploring in detail. Molar desire has several parts: 1: an implicit confidence in the coherence of a world (that what is desired will fit into a world in a positively valanced way), 2: some (often vague, intuitive, or even mistaken) sense of how what is desired will fit into or alter that world, and 3: an impulse to pursue what is desired that feels consonant with oneself, one's values, and one's projects. It is a drive for connections and to make something happen. These result from the successful organization of molecular desire. Personalized algorithms offload much of this. They curate and maintain access to a world for you (e.g., my knowledge of professional skateboarders has decreased relative to the increase in ease at which social media provides me videos of them). They develop profiles of likes and dislikes, past views, how long you lingered, etc. This effectively organizes one's desires for them. If so, then this weakens one's sense of the coherence of a world. It deskills the imagination that otherwise must produce a sense of how what is desired will cohere with or change that world. These algorithms can provide what one would want even before the impulse arises, thereby short-circuiting the reward of reinforcement learning, since the impulse is not first stoked. When the impulse is stoked, if the result is disappointment or only slight enjoyment before being distracted elsewhere, the reward signal is also weak. These cases only proliferate as markets are flooded with low-quality goods and generic content, i.e., slop. In short, the proliferation of such algorithms and the slop meant to feed them risks offloading our capacities to organize molar desire. This is particularly distressing as those capacities are central to the very process of self-formation.

Desire is also a building block of social groups. With only molecular desire, spontaneous organizations can be generated, but they fall away just as easily as they form. It is only through the organization of desire into molar group desire that such groups can maintain a structure of their own and so persist beyond the individuals who make them up. This is unlikely to be done without organized subjects who take this on as an active project, who make the group desire an integral part of their own molar desires. The same is true of reshaping existing social structures. This is also the ambivalent role of the pervert as the one who invests desires elsewhere than its current social arrangements. In doing so, they make it possible for others to desire in new ways as well. However, this works through a connection with molecular desire and so introduces an anarchic element to that arrangement.

Molecular, that is, unorganized, desire is anarchic and precedes the subject. If so, then molecular desires are not "yours" as they precede you as such. That we feel impulses that are strange to us, that we do not recognize or understand why we feel compelled by molecular desires is normal and a potential source of new joy, even if the strangeness is exacerbated in the age of advertising. However, control threatens molar desire itself. Without a molar organization of desire, one becomes unable to desire one's own desires, as Stiegler puts it. This is not a lack of desire as such, but an inability to make molecular desires one's own, to select among them, to integrate them into a self, and to feel the impulse of them in a harmonious way. Deleuze and Guattari refer to this state as the catatonic BwO, the individual whose sheer excessive and contradictory potentialities render it completely unable to act or be affected, (Deleuze and Guattari 1987, 150). Everything circulates but nothing moves. With this enormous excess of potentialities disconnected from projects, lasting social ties, desire for others' desires, and the like, no constellation of habits or goals can last long. Choice paralysis is but one particularly common manifestation of this situation under control. Taken to its limit, however, the subject is too disorganized to function as a recording surface, and so they cannot identify with what they do or with what happens to them.

To further diagnose this, we must note here how lack is only a possible effect of desire, not its source or condition. One understands oneself as lacking something only after there is a world and an impulse for something. I desire the coffee both when I am out of coffee and when I am sipping it, but only in the former case do I encounter what I desire as a lack. That this misunderstanding of desire is the target of Deleuze and Guattari's critique in *Anti-Oedipus* is well known. They diagnose it from Plato to Lacan. But the misunderstanding continues. It is defended both in everyday and academic discourse. We are told we have too much, everything is too easy, and this excess supposedly undermines character formation. We saw a version of it from Borgmann in Chap. 2, though Han is the most recent defender of this position under what he calls negativity. On this misunderstanding of desire, the cure is to embrace lack, to self-flagellate, to seek and celebrate the frustration of our desires. As if our problems could be solved by hollowing our lives out more, by embracing ever more asceticism. Even the other must be encountered not as an excess over us but as an inaccessible lack on Han's account, (see Han 2017).

When we reject this dominant account of desire as lack, what are we left with? If not negativity, if not a lack of lack, then what produces this suffering from excess? In short, the disruption of heteronomy. One forms themselves in response to others. In anticipating the needs and desires of others, we mediate our capacities, our molecular desires, and the like through the constraints of the world and the needs of others. This gives them shape. We have not just the various roles and identities society offers (which we may take up, reject, or innovate on), but also the demands of the other that provide some potential selection and often the impulse to novelty. This is obvious when we think of how our personalities modulate and change when moving between different friend groups or different social settings. Heteronomy then makes the organization of desire possible. Technics plays an important role here as well in that it provides a storehouse of possible lives, possible ways of being. Certain artifacts (musical instruments, factory equipment, work scheduling software, anything that needs ongoing maintenance, etc.) demand that we shape ourselves in response to them. All this nurtures and provides structure to our worlds, to the imagination required to organize desire. In responding to heteronomy and mediating this response through artifacts, environments, others, etc., we undertake the process of organizing desire as such, of building paths and connections between people, places, and things that channel our energy into the maintenance of psychic and social structures. Embracing lack would only harm molar desire further.

4.5 Truth

Truth makes little sense from the point of view of an individual subject alone, and less so for the lack of any subjectivity at all. For the individual subject alone, what matters is efficacy, that what one does satisfies one's desires (and here the point stands even if we pretend one could have molar desires without others). Whether what I believe is true will not be a concern as long as these beliefs allow me to achieve my aims (hence why pragmatism so often bottoms out in an empty, Rortyan bourgeois liberalism). The concern for truth arises, as Levinas argues, only in response to the other and their demands on me (Levinas 1969, 102). A deep concern for the truth is inherently social, even if truth, like desire, is fundamentally threatening to specific social identities. In the coming of the other, I must account for myself and my actions. In doing so, the whole of truth is brought into our sphere of concern. It is not just the particulars of the situation, but also the nature of reality that becomes a problem in our accounting of ourselves. In responding to the demands of the other, we determine facts of the local situation, seek the scientific accounts that are relevant, even pose perhaps unanswerable metaphysical questions which would nonetheless change the accounting (e.g., free will vs determinism). If there is no self to which the other may come, if there is just a roving bundle of inchoate desires, then truth is an empty word. Wonder and curiosity may drive my taking pleasure in the things of the world (though we saw with desire what little of the world remains without heteronomy), but it is concern for others that transmutes curiosity and wonder into truth.

A sort of epistemic substitution is at work here, parallel to ethical substitution but with a slightly different structure. The question of truth asks what would anyone put in my place conclude? What could they learn and what would they decide? But also: was I wrong? Was I justified? Without heteronomy, the separation between subjective, intersubjective, and objective cannot be made. This is why there is such a strong impulse to share (what we take to be) the truth. To say that a concern for truth arises from connection with others, it is both a justification of one's self and actions, and a (sometimes misguided) attempt to care for others to try to convince them, to teach them.

Control is adversarial to truth in that information is faster than truth, (Han 2022, 24). The lack of an interval of reflection, the infinite modulation of our connection to others or our subjection to the faceless mass of the online other undermines the possibility of coming to account. Control weaponizes the call to account through continual confession, which disrupts the internalizing movement of reflection. Without this, information cannot be processed, correlated, and made into knowledge, into what would allow one to speak the truth in any serious sense. Internalization here does not make information subjective, but rather is the work of synthetizing, of preparing it to be shared. "Interiorization is a going beyond oneself; it is universalization," (Hadot 1995, 211). The loss of heteronomy, either in the faceless encounter with the mass or lack of a call to make one's beliefs sharable, then hollows out any meaning to truth as such.

4.6 Conclusion

While I am sure there are more goods that depend on interiority, it is those explored above that I think are the most important to survive the poststructuralist critique of the subject. Interiority is the condition of a deep heteronomy, and it is productive. It is the answer to Deleuze's question, again, in his *Foucault* book, "is there an inside that lies deeper than any internal world?" (Deleuze 1988, 96). Interiority may be the condition of a self or subject and its internal world, but it is much more than that. At least three fundamental things do not make sense without an asymmetrical connection between one interiority and another: ethics, molar desire, and truth. In short, ethics requires an other whose interiority I cannot know fully and so must be receptive to them beyond my understanding of the world. Desire requires both that I maintain an interiority and that the interiority of the other cannot be subsumed into me. And finally, truth only makes sense if we are called to explain ourselves to another that we are in an ethical relation with.

If so, then Surveillance Capitalism is by its confessional structure inherently opposed to ethics, desire, and truth. By emptying out interiority through transparency, the very conditions of sociality are undermined. Without a robust interiority, without one's own excess that stands out from the world and from others, ethics, truth, and desire lose all meaning. The self *qua* the cop in your head may not be worth saving, but interiority as the condition of these three great goods is. Fleeing back to the self for the sake of individualism and autonomy is reactionary, but a resistance to surveillance capitalism and control in the name of interiority can and should be done for the sake of difference and otherness as well as for our selves. Only this can maintain some of our fundamentally important capacities to the degree required to respond to our current situation. "Since the workers are not destroyed as human beings, but are simply utilized in inhuman ways, their critical, intelligent, conceptual faculties, no matter how deadened or diminished, always remain in some degree a threat to capital," (Braverman 1998, 96). The struggle over interiority is the struggle to maintain our capacity to be this threat.

It is worth repeating that the goods argued for above are not the only or the highest goods. They are essential parts of a good life, and they are worth affirming here so that we do not fall into common traps of post-structuralist excesses or misreadings (such as those who saw *A Thousand Plateaus* as a retreat from the radicality of *Anti-Oedipus*). But these goods are not unqualified. Each of them can be taken to a dangerous degree such that they become inimical to life. We have seen how truth taken to its limit in confessional transparency undermines interiority itself. There is a need for guile, secrecy, and cunning in the good life, a necessity of privacy, opacity, and fabulation as much as an inherent good to the truth. Similarly, our receptiveness to others can become our domination easily. And there is no virtue to be found in an unwavering identification with one's own projects, goals, and values; successfully organized molar desire that becomes unresponsive to all demands for reorganization is vicious. In short, remaining who and what we are is not a good in itself.

Our flourishing depends not only on our metastabilities, but our individuations as well. There are unique goods of the disruption of heteronomy and of the temporary submersion into the transindividual (both spontaneously, in that collective feeling of undergoing a powerful event together, and as a practice of group-cultivation). Breaking down one's own subjectivity to reforge it anew is often demanded by ethics and required to see through the traps of confusing ethics with parochial morality. Learning destabilizes the subject as much as it challenges what is taken as true, without which we could not separate knowledge from social convention, nor objectivity from intersubjectivity. All of this is possible only because underneath our subjectivities, our groups, our values, our worlds, and our projects, is the anarchy of molecular desire. The source from which subjectivity flows is also always pushing us beyond ourselves towards other people, animals, ideas, flows, towards couplings across the machinic phylum, forever operating via a perversity without law. Without this, the good is not possible, and yet so much of human morality is formed in the fear of this fact; the harms resultant of this fear are beyond reckoning. These goods will take a more prominent role as we turn towards the positive project of this work.

For Levinas, the other is generally human. For Deleuze, anything could potentially trigger this encounter, though he focuses on how such encounters can submerge the subject back into the level of individuation. There is then a question of whether there is heteronomy, and so ethics, in relation to non-humans. I think there is. Insofar as everything has some capacities (virtualities) which it can only show indirectly through its manifestations (actualities), everything has some depths in excess of what we try to make of it, even if it does not have the psychic interiorities that Levinas privileges. These metaphysical but not psychological depths of things may also be engines of sublime encounters that break down our stabilities and throw us deeper into transindividual processes. However, I see no reason why cases of intense care of non-human others could not also be a relation of heteronomy. Whether this is parasitic on these structures of heteronomy having been produced through encounters with the human other, I do not know, though I am skeptical. Instead, I take it that different sorts of interiorities come with different demands, different features, etc., but may all potentially be heteronomous. We may substitute the good of a work of art or a forest in place of our own needs. Anyone who has repaired a car knows the demand for truth that it makes on us as we try to diagnose the problem. Importantly, our capacity to organize our desires is always mediated through non-humans, as we will see with the discussion of ritual in the next chapter.

References

Braverman, Harry. 1998. *Labor and Monopoly Capital: The Degradation of Work in the Twentieth Century*. 25th Anniversary ed. Monthly Review Press.

Cohen, Alix. 2008. Kant on Anthropology and Alienology: The Opacity of Human Motivation and Its Anthropological Implications. *Kantian Review* 13 (2): 85–106. https://doi.org/10.1017/s1369415400001242.

Deleuze, Gilles. 1994. *Difference and Repetition*. Translated by Paul Patton. Columbia University Press.
Deleuze, Gilles. 1988. *Foucault*. Translated by Sean Hand. University of Minnesota Press.
Deleuze, Gilles, and Felix Guattari. 1983. *Anti-Oedipus: Capitalism and Schizophenia, Vol 1*. Translated by Robert Hurley, Mark Seem, and Helen R. Lane. University of Minnesota Press.
Deleuze, Gilles, and Felix Guattari. 1987. *A Thousand Plateaus: Capitalism and Schizophrenia, Vol 2*. Translated by Brian Massumi. University of Minnesota Press.
Foucault, Michel. 2012. *The Archaeology of Knowledge: And the Discourse on Language*. Vintage.
Gamez, Patrick. 2022. A Friendly Critique of Levinasian Machine Ethics. *The Southern Journal of Philosophy* 60 (1): 118–149. https://doi.org/10.1111/sjp.12455.
Gunkel, David J. 2018. *Robot Rights*. The MIT Press.
Hadot, Pierre. 1995. *Philosophy as a Way of Life: Spiritual Exercises from Socrates to Foucault*. Translated by Arnold Davidson. Wiley-Blackwell.
Han, Byung-Chul. 2017. *The Agony of Eros*. The MIT Press.
Han, Byun-Chul. 2022. *Infocracy: Digitization and the Crisis of Democracy*. Polity.
Howells, Christina, and Gerald Moore, eds. 2013. *Stiegler and Technics*. Edinburgh University Press.
Kant, Immanuel. 2006. *Anthropology from a Pragmatic Point of View*. Translated by Robert B. Louden and Manfred Kuehn. Cambridge University Press.
Kronsted, Christian. 2025. Collective Effervescence as Self-Organization and Enaction. *Journal of Social Ontology* 11 (1): 1–27. https://doi.org/10.25365/jso-2025-8732.
Levinas, Emmanuel. 1969. *Totality and Infinity*. Translated by Alphonso Lingis. Duquesne University Press.
Levinas, Emmanuel. 1998. *Otherwise Than Being: Or Beyond Essence*. Duquesne University Press.
Métais, Fabrice, and Mario Villalobos. 2022. Levinas' Otherness: An Ethical Dimension for Enactive Sociality. *Topoi* 41 (2): 327–339. https://doi.org/10.1007/s11245-021-09772-z.
Paul, L. A. 2016. *Transformative Experience*. Reprint ed. Oxford University Press.
Schreel, Louis. 2024. *Deleuze and the Immanent Sublime: Idea and Individuation*. Bloomsbury Academic.
Simondon, Gilbert. 2020. *Individuation in Light of Notions of Form and Information*. Translated by Taylor Adkins. Vol. 1. University of Minnesota Press.
Sparrow, Tom. 2013. *Levinas Unhinged*. Zer0 Books.
Stiegler, Bernard. 2010. *For a New Critique of Political Economy*. Translated by Daniel Ross. Polity.
Striano, Francesco. 2024. Towards a Transindividual Ethics for the Dividual Condition: A Program. *Chiasmi International* 26 (July): 95–110. https://doi.org/10.5840/chiasmi20242610.
Thornton, Edward. 2025. *Deleuze and Guattari's Philosophy of Transformation: Individuating the Line of Flight*. Edinburgh University Press.
Zuk, Peter. 2024. Mental Integrity, Autonomy, and Fundamental Interests. Extended Essay. *Journal of Medical Ethics* 50 (10): 676–683. https://doi.org/10.1136/jme-2023-109732.

Chapter 5
Subjectivation

> *"Intense pleasure in skill, machine skill, ceases to be a sin,"*
>
> *—Donna Haraway, Cyborg Manifesto*

While we have defended aspects of interiority worth saving from control despite the post-structuralist critique of the subject, these aspects cannot be saved at the level of willed action by the subject.[1] Control and the technologies that make it possible work on the level of individuation processes. This metaphysical fact makes several prominent ethical theories implausible for guiding us towards a good life. Deontology requires the fully formed, self-legislating, autonomous subject. This disregards how such features, insofar as they are desirable at all, are conditioned by a prior heteronomy that is under threat by control. Utilitarianism has a similar flaw but in regards to desires. It takes the subject to be an atomic collection of desires. Yet if our desires are shapeable, if we can lose or gain desires depending on what happens to us, often outside of our choices, utilitarian calculus breaks down. How could one calculate the greatest happiness of the greatest number of people when one accepts the mutability of desire, the possibility of radical transformation, and the good of the opacity of the other? Which desires should you embrace? In deciding what possible version of you, should your calculations privilege your current desires or future desires? Would it be better for you to acquire a sexual fetish you currently lack? These are all possible yet incalculable. More importantly, to model the opacity of the other is to attempt to undermine it.

Of the major ethical traditions, only care and virtue ethics may survive this situation. For virtue ethics, this is because it focuses on questions of how to shape ourselves well; for care ethics, it is because of its attention to our care for others through their opacity. In an age of constant information fluctuation, of constant belief modulation, we cannot depend on beliefs, rules, or calculations alone. Given the importance of heteronomy, the attention to and care of our relations to others is

[1] This chapter includes and greatly expands on material originally published in Ardoline 2022.

M. J. Ardoline, *Interiority Against Control*, Sophia Studies in Cross-cultural Philosophy of Traditions and Cultures 47,
https://doi.org/10.1007/978-3-032-16245-8_5

necessary to a good life. Since heteronomy requires a subject capable of maintaining their metastability, the cultivation of virtuous dispositions is equally necessary. Here is a further advantage of the virtue tradition, which never believed in belief alone: the character of the person modifies and qualifies the belief. There is little good to be found in those who are dedicated to non-violence only because they are cowards.

In *Society Must be Defended*, Foucault looks for ways of opposing biopower, and he suggests that it may be possible to reactivate the notion of Right central to sovereign power against the Norms of biopower. To do so, Right would need to be stripped of its dominating characteristics. We see this maneuver in how liberalism defends rights in a way that aims to make us tiny sovereigns of our home or our body. This can then be extrapolated to control in that the normative force of biopower could be taken up as a weapon. For example, as compensation for control, you may now impose medical norms on yourself: you may embrace your gym routine, you may seek opposition in "alternative" medicine grifts, you may self-diagnose thereby gaining some power by pathologizing yourself. Much of this is worrisome for its obviously reactionary aspect which embraces the dredges of a fading form of power to oppose the new one. However, it is not without its merits, and this uptake of discipline and norm from below is what Foucault sees as interesting in the technologies of the self and the care of the self, (Foucault 2015, 26). If so, then these techniques must be stripped of their more odious elements.

Here our rejection of an ethics based on knowledge alone may be somewhat helpful in separating the reactionary from the liberatory, though this requires rejecting Foucault's specific notion of care. Fuggle 2013 notes the difference between care in Foucault and Stiegler. For Foucault, it is *souci*, a French world meaning concern with or to worry about something. It connotes a concern with knowledge or truth. The care of the self in this sense (paired with the Delphic command to know thyself) is a type of self-knowledge tied to a project of ethical becoming.[2] Foucault's notion of care, then, is too indebted to the same logic that associates light, knowledge, and the good that control is built upon. When discussing care, Stiegler instead uses *soin*, which more directly connotes taking care of or caring for something. This sort of care is more active, less reflective. Central to this activity of care is attention, which is always mediating the individual and the group. Attention is both focusing on (psychic) and attending to (social, in the sense of caring for), (Stiegler 2015, 152). This is, I think, key to any ethics that seeks the good life under control and the techniques of ubiquitous, non-disciplinary surveillance on which it depends. It is in explicating what such care for our self and others would be if it were broadened beyond this simple equation of knowledge and the good that must take us outside the continental tradition, and the so-called western traditions more generally.

[2] See Cerrato 2025, 88–92 on the relationship between the care of the self, self-knowledge, and how this motivates Foucault's turn to *parrhesia* as possible ethical reconciliation between truth and power.

5.1 Reconsidering Ritual

We have argued that the post-structuralists provide powerful diagnoses and critiques of the present situation, but little in the way of positive projects for how to live well despite and in resistance to them. James Garrison proposes a fascinating and original response to these lacks in continental thought: a symbiosis between Continental thought and the Confucian tradition.[3] At first blush, this may seem to be a strange pairing, but Garrison makes a convincing argument that the two traditions have much in common. Both traditions have robust accounts of subjectivation, that is, of how social forces form subjects, in which the subject is understood as fundamentally relational, discursive, embodied, and ritually impelled, (Garrison 2021). We should briefly gloss this latter category as it is central to Confucianism generally. Ritual names the interrelated system of all the social practices, rules of etiquette, norms of governance, regimes of training, etc. that structure the various roles of a society. Ritual is as minor as table etiquette and as major as the norms of international relations. What Foucault calls discipline is a subset of ritual in the Confucian sense. However, Confucian ethics sees the proper institution of ritual practice as fundamentally necessary for good people and good societies. This is because ritual is understood to be the proper ordering of the world which when followed correctly (i.e., when one does not just go through the motions) shapes subjects into good people and good practitioners of their various social roles. Following ritual also means performing maintenance on a good society and ensuring its harmony with the natural world, assuming we have gotten the rituals correct. We will see below that the Confucians differ on how ritual is to be understood, and our synthesis in this chapter will end with an account of ritual mediated through Deleuze's metaphysics to purge the conservativism inherent in the Confucian notion.

Furthermore, in both Confucian and Continental traditions, the subject is not an isolated, monadic individual. Rather, it is the effect of social forces, the languages one is taught and by which one is shaped, the body one finds oneself in, and the accumulation of practices one engages in and undergoes. Furthermore, both traditions agree that these subject-forming forces must be understood both in their aesthetic dimensions (how one is receptive to their world, what affects they possess, and how they evaluate and get evaluated) as well as in terms of the techniques subjects are trained in and have applied to them.

As many of these positions are seen as either unique to or at least distinctive of the post-structuralist and critical theory strands of continental thought, it should be shocking and fascinating for one steeped in these intellectual lineages to find such similar ideas in a context so distant in time and space. More to the point, Garrison

[3] I should say that this is novel for continentalists and especially anglophone thinkers in this tradition, and for the specific thinkers proposed, but not for comparative philosophy generally. There has been important work on encounters between Eastern and Western philosophical traditions in Eastern contexts throughout of the twentieth Century, especially in work by Chinese Marxists and the Kyoto School.

argues that "having had such a long history in which to develop its own terms, Confucianism can address the conjunctions of ethics, aesthetics, and politics that occur in person-making in ways that the best, though still ultimately tradition-bound and reactive efforts from Euro-American critical theory cannot," (Garrison 2021, 1). Given these common commitments, the Confucian tradition must have much to offer to the significantly younger Continental tradition, even if as we will see, the latter should not adopt the former's insights wholesale.

In making this connection, Garrison is aiming to ameliorate one of continental thought's most glaring issues: the ambivalence of subjectivation. This ambivalence results from the main ethical and political question raised by subjectivation: if we are products of our social realities down even to our bodily comportment, speech, and unconscious desires, from where is something better to be produced? The answer, generally, is resistance, yet this is quickly followed by the caveats that the capacities and often desires for resistance are formed by precisely the structures one sets out to resist. Like an electrical circuit, most resistance is a part of the flow of power in these structures, not opposed to it. Whatever this worldview's virtues are (and it has many), a feeling of futility is a reasonable effect of holding it.

Garrison does not reject either the concepts or practices of resistance found in the continental tradition. Rather, he argues the Confucian tradition offers a new avenue to continental thought via its understanding of autonomy as a result of subjectivation itself (Garrison 2021, 19). Garrison is humble about his contribution which "is meant to supplement rather than supplant resistance strategies," (Garrison 2021, 19). While I agree that resistance strategies should not be devalued or supplanted, I think this project contains the promise of a new future for the continental tradition. In showing that the accounts of the subject one finds in these two traditions share so much, one is forced to reckon with the fact that subjectivation is precisely the source of Confucian optimism. Though they disagree on how, the great Confucian thinkers agree that it is because we as humans are shaped by social relations, power, aesthetic capacities, and techniques in all aspects of our being that the problems we face are surmountable by human ingenuity, especially when these problems are social in nature. In fact, if one accepts subjectivation in general, there is no other source from which the conditions of individual autonomy (i.e., interiority and heteronomy) and the ability to alter our world could spring. This alternative ought to be jarring to the continental thinker who sees subjectivation precisely as a reason to doubt or heavily curtail the possibility of human autonomy.[4]

We should wonder why there is such a difference in the appraisal of subjectivation between the two traditions. Giving a complete explanation would be a full study in its own right, but it is worth staking a reason to think that the continental tradition's pessimism is unwarranted. This is doubly so if that reason is stronger than the appeal to tradition that Confucianism has been around longer, or the mere

[4] One could respond that the promise of a positive reconciling of autonomy and subjectivation has always been in the strands of continental philosophy that embrace Spinoza. Again, I would agree, but add then that an encounter with the Confucians can allow us to actually make good on that promise.

hope that something better than circumscribed resistance is possible. I would stake the claim that we should give more credence to Confucian autonomy than critical pessimism because the latter has had to deal with the fallout of losing more a robust tradition of autonomy. For example, we see from the *Nicomachean Ethics* onward an equation between the Good and the notion of self-sufficiency (Aristotle 2019, I.7). Barring a few important exceptions, this is distinctive of the Euro-American traditions through to today. Nor is it merely a leftover Aristotlean imprint as this tradition came to increasingly embrace this equation more fully than Aristotle himself did (for example, in Christianity's doctrine of individual salvation coupled with Augustine's idea of free will, Stoicism's carving up of the world into what we can control and what we cannot, or the American transcendentalist's value of rugged individualism). Its culmination is perhaps Kant's account of noumenal autonomy beyond the bounds of the phenomenal world. Critical theory and the continental tradition more generally are an inheritance of and reaction against this history, where the account of subjectivation puts the lie to the belief in the self-sufficient individual. And yet, even if one agrees that these criticisms are true, one formed in these traditions often cannot help but feeling the rejection of self-sufficiency as a loss of a good, if not the Good. Perhaps this is why post-structuralist thinkers continually deal with charges of relativism and nihilism from their opponents who, knowingly or unknowingly, retain the equation between the self-sufficient and the Good. Continental thought may have banished the figure of the self-sufficient, autonomous individual separated from their environment and sociality of previous Euro-American traditions, but it has not completely overcome the desire to be such an individual. Hence the pessimism.

For the Confucian tradition, no such loss has taken place. This is not to repeat the stereotypical oversimplification that "western culture is individualistic; eastern culture is collectivist."[5] There are important individualists in Chinese philosophical history. The most important of such thinkers whose ideas survive in text to today is perhaps Zhuangzi, who is obviously no small influence on (again, so-called) Eastern traditions. The ancient Confucians respond to such individualist thinkers throughout their texts, with Mengzi being particularly concerned by the popularity of the ethical egoism of the Yangists, (a school whose texts are now lost to us, but who likely inspired Zhuangzi to some degree). However, such individualist positions are the Confucians' contemporary rivals; they do not constitute a historical orthodoxy that Confucianism is a reaction to in the way that the continental tradition is reacting to its Aristotelian history. In that sense, there is no loss of a perceived good that comes along with accepting the view of the subject as socially, discursively, bodily, and ritually produced. This socio-historical fact seems to me to be good reason to challenge critical pessimism and the return to the Greeks seen in Foucault's later care of the self project. This is not to say that continental thought or Western traditions more generally cannot overcome their historical equation of the Good and

[5] See for example MacIntyre 2007 or Machek 2025, 26–7 for how ancient Greco-Roman ethics place greater emphasis on social role than is commonly appreciated, especially given how concerns about the individual soul come to dominate the later tradition.

self-sufficiency, but that it will likely be easier to do so through an encounter with a tradition that does not have the same historical baggage.

In this vein, I will note before moving on that there is of course much in the Confucian tradition (especially the ancients) that is to be rejected out of hand by those in the continental tradition. For example, the unquestioned equation of order with Good and disorder with Bad common to many ancient traditions is especially pronounced in Confucianism. Furthermore, this leads Confucian thinkers to understand hierarchy qua order as inherently good and justified. Confucian thought's sanguine appraisal of hierarchy, order, and the state's role in encouraging and enforcing morality often appear to the modern reader as (and in some cases, are) naïve. Continental thought's critical bent, especially regarding the abuses made possible by norms, institutions, and subjectivation themselves, can offer a useful corrective. Equally, the equation of Good and order is to be rejected as chaos is necessary to the formation of structure in the first place as well as its change and alteration. Here it is notable that ancient Confucian accounts of subjectivation do not have a robust metaphysics of individuation underneath them. Providing one is central to the synthesis below. This equation of order and the good should also be rejected as it is a central tenant of control: "Cybernetics, that science-as-steersman, made an angel of control and a devil of disorder. But perhaps disorganization, noise, and uncontrollability are not the greatest disasters to befall us. Perhaps our calamities are built largely from our efforts at superorganization, silence, and control," (Galison 1994, 266).

Similarly, Continental thinkers have good reason to critique Confucian values such as *ren* (often translated as "humaneness" or "benevolence") and filial piety as too anthropocentric, too humanistic, too naturalized, and too conservative. This is, I take it, not a reason to reject the symbiosis Garrison posits, but rather an argument in favor of it. If both traditions agreed on subjectivation but the Confucians had useful practical advice and the continental tradition had no good critiques thereof, then there would be little more to say than that the Confucians are right, or at least more useful. Fortunately for us continentalists, this is not the case. The point is not to become Confucians today, but to seek something new.

To that end, I will for the most part leave these critiques of Confucianism to future work and instead focus on producing an encounter between the two traditions. Here we break again with Heidegger. He claims that because the current world technological crisis of reducing all things to their relation in a technological system, all beings to their potential as standing reserves of energy, and all values to efficiency is the result of Western metaphysics, it is therefore only from within the Western tradition that its overcoming can be found. When asked if something like Zen Buddhism may be helpful in our present situation, Heidegger responds that "Thought will be transformed only through thought that has the same origin and determination," (Heidegger 2017, 62). This is a laughable claim, and one based on a view of cultural supremacy. Heidegger thinks that the German language has a special relationship to Greek thought and language, and since technology as *Gestell* results from Greek thinking, it can only be responded to and overcome by something internal to or continuous with it, something capable of returning to the origin.

It is as if to overcome their character flaws, one believed they had to return to the womb. It is probably easier, and less Oedipal, to make new connections.

Furthermore, if one is going to privilege origins and the equation between language and thought (I reject both), then there are several note-worthy features of this encounter that even a Heideggerian would have to admit. First, ancient Confucianism precedes the contact with Buddhism, and so precedes the synthesis of thinking from languages with Indo-European roots like Sanskrit. If one cares about such things, then an intellectual tradition whose origins are free of the reifying legacy of the verb "to be" carried on from Proto-Indo-European onwards should be of interest. Second, there is an alternate relation to technics, or rather the lack thereof as Hui 2016 argues. Stiegler claims that at the birth of Western philosophy, *techne* is separated from *episteme*, and the former gets repressed. Hui argues that no such separation takes place in the Chinese tradition. Prometheus, for example, is not a universal myth.[6]

In more recent work, Hui has sought a sort of Heideggerian multi-cultural approach to these concerns, (Hui 2024, 248). There he argues for epistemic diversity that broadly tracks current major cultural traditions and division. While he embraces a metaphysics of individuation as fundamental to this, he opposes his project to those he sees as forms of creolization and symbiosis. Here I think Hui betrays his own metaphysics in that he rejects the power of difference and disparity to produce something new. This is seen straightforwardly in his critique of creolization. Creolization, he argues, is effectively additive. That is, it is just adding parts together alongside each other in a way that does not produce new forms. Here he gives the example of a milk, tea, and coffee drink made popular in Hong Kong as a "creolization" of East and West that amounts to little more than putting two separate things together. This is a deeply short-sighted view of what creolization or symbiosis may produce. For example, this is not how a creole language works. In blending languages together, a creole language does not merely consist of words from multiple languages; it can generate its own idioms and grammar, that is, a new structure. Some of these structures are at first unintelligible to speakers only fluent in the parent languages. This is surely not a merely additive process. In critiquing Hui, I don't consider my project a form of Creolizing in the sense of on-going projects in work such as Sealey 2020, Monahan 2011, and so on, though this critique is a defense of those projects as well. Rather, Hui's critique is wider than those that explicitly take up the term; it is seemingly aimed at any cross-cultural philosophical symbiosis.

If Hui is right that the Chinese tradition lacks a technics (Hui 2016), it is not because it lacks thinkers of technology. Rather, it is because those thinkers did not become cornerstones of the tradition. In particular, Mozi and Xunzi are both

[6] This raises the question of whether the repression of technics is a universal phenomenon, or one that results from a certain tradition. I tend to think it is at least wider than its supposed Greek origin in that notions such as those in the Vedic traditions like *atma,* in Buddhism like storehouse consciousness, and possibly *Tian* in many of the Warring States era schools (as the preserve of the proper order of human integration into nature) all spiritualize and repress the fundamental role played by technology. In the latter case, we will see below that Xunzi is an exception.

thinkers who are marginalized within their tradition because of their opposition to some of its fundamental positions. Surprisingly, both are also profound thinkers of technology. This makes an encounter with them untimely in Deleuze's sense. Xunzi will be the focus of this chapter, Mozi of Chap. 7. In rejecting Heidegger's restriction to traditions and their origins, and in recovering untimely thinkers in the traditions being brought together, it is my hope that this synthesis escapes both the common clichés of the East/West divide and achieves a full synthesis rather than a mere addition.[7]

5.2 Mengzi vs. Xunzi

In setting out his synthesis, Garrison argues that of the ancient Confucians, Mengzi should be preferred to Xunzi for projects which seek a positive, potentially optimistic understanding of subjectivation (Garrison 2021, 176–180). In response, I will make a case for the advantages Xunzi's account offers specifically to our understanding of subjectivation. This is not to reject Mengzi's usefulness to Continental thought, nor Garrison's particular constellation of Butler-Foucault-Mengzi. Instead, I offer a supplement constellation of Deleuze-(and Guattari)-Xunzi.[8] In doing so, I will argue that adding Deleuze's ontology to Xunzi's anthropology and his constructivist[9] account of the *Dao* (Way) provides a useful understanding for how to experiment with new ways of living such that we can become an active part of our

[7] I imagine a reader familiar with both Deleuze and the Warring States era schools might wonder why the synthesis is with the Confucians rather than the Daoists. There are two reasons. First, given certain fundamental commonalities between Deleuze and the early Daoists, this seems to me to be less a synthesis and more of an intellectual inbreeding. A likely result of such a union would be to push Deleuze's metaphysics closer to his process-only readers, and this would undermine several key arguments of this work. Second, the *Daodejing* proposes that a good leader should keep a population deskilled to a degree that keeps them happy and easier to manage. As deskilling is a fundamental danger of control, the early Daoists seem to me ill-equipped to respond. Here, however, the skill stories in the Zhuangzi are a potential counterpoint, though these often depict effortless skillful engagements in contrast with interiority.

[8] To my knowledge, the only work in English that puts Deleuze and Guattari and Xunzi into conversation is Yin 2013. Here, Yin contrasts Deleuze and Guattari's concept of becoming-Animal against the traditional Confucian and European distinctions between human and animal, relative to the novel Wolf Totem. While I agree that these thinkers are opposed on that account, I aim to show a unique affinity between them. Ott 2024 provides a robust encounter between Deleuze and Neo-Confucianism, but as I am restricting this encounter to Warring States era Confucianism, I do not engage with this work.

[9] A fundamental debate within Xunzi scholarship is whether Xunzi is a realist or a constructivist about the *Dao*. It would be beyond the scope of this paper to adjudicate the debate. Instead, I make use of Kurtis Hagen's reading of Xunzi, specifically Hagen 2002, 2003, 2005, 2007, for the sake of setting up this encounter with Deleuze. For an example of a realist alternative to Hagen's view, see Ivanhoe 1991.

own subjectivation, as well as that of others. Moreover, it will show critical pessimism about subjectivation to be mistaken.

While Garrison argues for a productive encounter between the two traditions generally, he argues that embracing Mengzi's Confucianism over that of Xunzi's is preferrable for the continental tradition, (Garrison 2021, 174–184).[10] While complex and varied, the opposition between Mengzi and Xunzi is often succinctly characterized by their disagreement about human nature. Mengzi equates human nature with our potential for goodness, whereas Xunzi sees humans by nature as a chaotic and somewhat detestable bundle of desires. Garrison's argument begins here. Two major points of Garrison's argument for the preference of Mengzi over Xunzi on human nature are that 1) Mengzi's view of human nature as always potentially good provides a better starting point for an optimistic view of subjectivation than Xunzi's authoritarian harshness, and 2) Xunzi maintains an ontologically flawed distinction between the internal and the external. Garrison does not reject Xunzi completely. Instead, he gives a limited endorsement of Xunzi's defense of and insight into the usefulness of ritual in shaping subjects. Yet, this endorsement comes with a steep condition. Garrison councils that we should separate Xunzi's defense of ritual and practical advice from his account of human nature (Garrison 2021, 180).[11] I will provide arguments against both the optimism and internal/external challenges as a way of developing Xunzi's usefulness for Continental philosophy and for showing what Deleuze's ontology can both offer and make use of in Xunzi.[12]

Beginning with 1), the human nature debate between Mengzi and Xunzi is the most discussed debate in studies of the ancient Confucian thinkers, and I have little new to say about it. However, it is likely unfamiliar to many continental scholars, so I will briefly recount the debate. This debate is generally glossed as an argument over the moral status of human nature, with Mengzi claiming it is good and Xunzi claiming it is bad or evil, but this is a gross oversimplification. Van Norden describes it instead in terms of moral agency and cultivation (Van Norden 1992). With this

[10] Here, Garrison is in good company. For example, when Varela 1999 is looking for a thinker to help highlight the ethical dimensions of embodied cognition and phenomenology, he also turns to Mengzi. However, Garrison's use of Mengzi is worked out in much more detail.

[11] Garrison arrives at this limited endorsement through showing that Mengzi's and Xunzi's views converge when opposed to a common enemy, the Mohists. The Mohists opposed Confucian ritual, seeing it as extravagant and unnecessary, especially for laborers, and thereby a cause of poverty. They went so far as to condemn music (see chapters 7 and 32 of Mozi 2014). Xunzi responds that humans are shaped through their aesthetic senses, and so music, as well as ritual more generally, are required for human goodness to flourish. Garrison contends that such a defense does not depend on Xunzi's account of human nature, and that it could be just as easily paired with Mengzi's account (Garrison 2021, 180). We will make positive use of the Mohists in Chap. 7, and will highlight some commonalities between them and Xunzi that sets the latter apart from Mengzi.

[12] To be more specific, Garrison here argues for Mengzi in regard to Shusterman's project of Somaesthetics. Shusterman himself favors Xunzi over Mengzi. However, since Garrison is making use of the program of Somaesthetics as a part of ameliorated continental accounts of subjectivation generally, I am engaging with the issue at this more general level. For the sake of brevity, I will leave out direct discussion of Shusterman's Somaesthetics.

framing, we may understand it as a debate over the metaphysics needed to explain how best to produce virtuous people.

By saying human nature is good, Mengzi does not mean that humans are already inherently good or virtuous by nature alone. Rather, he tells us "As for what [human beings] are inherently, they can become good. This is what I mean by calling their nature good. As for becoming not good, this is not the fault of their potential,"(Mengzi 2008, 6A6.5–6). Here, the inherent potential towards goodness is the central point. Mengzi's view is that humans, by nature, have four "sprouts" that make morality possible: compassion, deference, disdain, and feelings of approval and disapproval (Mengzi 2008, 2A6.4–6). These sprouts are either nurtured or hampered by one's environment. They are then minimal capacities for the good that humans, in the right environment, will tend to develop as an internal tendency. If people are in an environment that provides what they need, and they have virtuous models to imitate (usually family members and local officials), then, Mengzi argues, these sprouts will flourish into the virtues of *ren*, shame, propriety, and wisdom, respectively.

Mengzi's phenomenological argument for the sprouts is famous. One sees a child who one does not know standing on the edge of a well. As the child walks along the edge, they stumble, falling towards the well. Mengzi claims that any human watching this event unfold would feel compassion and concern for the child, even if that feeling is a mere twinge (Mengzi 2008, 2A6). It does not matter if one acts on such an impulse; the existence of this feeling of concern is a sign of the four sprouts of virtues. That such moral impulses contain within them even a modicum of motivation to act on them means they are signs of the tendency to become good. We should also note that this concern, and compassion generally, is not necessarily empathetic; the child does not realize that they are in danger, (Ivanhoe 2017).

The sprout metaphor endures because it is particularly enlightening of Mengzi's position. He sees all humans as by nature containing basic faculties of virtue that guarantee us a source of potential goodness. This potential has the virtues as their natural tendencies. For example, if properly nurtured, the sprout of compassion will grow into the virtue of *ren*, the chief virtue of the Confucians. Like seeds, the four sprouts cannot be forced to grow; they can only be nurtured and given the right environment that allows them to flourish. By environment, Mengzi does not mean something like our contemporary notion of nature separated from humans. He means the natural world human life is enmeshed with but also the social arrangements that people find themselves in, the actions of their rulers, their family life, etc. So having a nurturing environment is not just having basic material needs met (though these are necessary conditions) but also living in a well-ordered society wherein most members behave in ritually appropriate ways and regard each other well. Promoting such an environment is, for Mengzi, how one follows the *Dao* (way or path), or proper moral way of life, and it is especially beneficial because it helps nurture others' innate goodness.

Similarly, if a person becomes wicked, it is not the fault of their nature, but their environment. On this view, we do not have countervailing inherent potentials to evil. There are no bad sprouts within us. Mengzi specifically argues against the doctrines that human nature is neutral (becoming whatever its environment makes of it) or that human nature contains tendencies to become either good or wicked, given one's environment. "In years of plenty, most young men are gentle; in years of poverty, most young men are violent. It is not that the potential Heaven confers on them [i.e., human nature] varies like this. They are like this because of what sinks and drowns their hearts," (Mengzi 2008, 6A7.1–2). Mengzi then gives the famous parable of Ox Mountain to illuminate his claim (Mengzi 2008, 6A8). Ox Mountain, he tells us, had all its trees felled for lumber and its vegetation harvested. When left alone, sprouts of new vegetation began to grow, but these sprouts were continually grazed on by livestock until Ox Mountain became bare. If one happened upon Ox Mountain, they would mistakenly believe it to be barren by nature, lacking even the potential to harbor vegetative life. "But could this be the nature of the mountain?" Mengzi asks. In other words, the mountain has no potential barren tendency within it, only the potential to grow vegetation. It is what happens to the mountain that prevents any growth. And despite it all, the sprouts still try to grow. According to Mengzi, such is the same for goodness in human nature even in the most wicked of people.

Xunzi, responding to Mengzi, argues instead that human nature is chaotic and detestable (Xunzi 2016, 248). While his position is often glossed in English as "human nature is evil," this is an oversimplification. First, a good bit of moral overcoding results from this English translation. With this poor summary, Xunzi's position takes on Manichean resonances that are of course not in the original text. When Xunzi describes in detail what he takes to be good and bad, he generally gives a distinction between the noble and the base, and between acting based on projects, goals, socials ends, or virtue versus acting on mere self-interest or partiality. Rather than evil, Xunzi argues we have haphazard and often inconsistent desires, and that we are by nature incoherent in ways that make it likely that we will do harm (hence, chaotic). This leads us to act in ways that ought to be despised by those who are properly cultivated. We may also do good things and even act on good motivation from time to time in our uncultivated state, but we will not be consistent; base and partial desires will usually win out. In short, Xunzi sees human beings as, by nature, a haphazard collection of desires and capacities with no innate normative tendencies. It is for that reason that our nature is detestable. In this, his account is structurally similar to Deleuze and Guattari on molecular and molar desire, but where the latter see productive tension between the two and positives to the molecular itself, Xunzi does not.

Important here is that Xunzi's view of human nature is not just normatively different, but also metaphysically different from Mengzi's. While Xunzi agrees that humans are capable of change and acquiring new traits, he claims that we do not have any inherent developmental tendencies, whether for good or bad (Xunzi 2016, 249). Xunzi's view here is similar to Aristotle's argument that virtue must be a habit because humans can acquire or lose them, and one cannot acquire or lose a nature (though Aristotle's account of potentiality relative to the virtues is closer to Mengzi). A thing's nature, humans included, is that aspect of the thing that is "the accomplishment of Heaven [*Tian*]"(Xunzi 2016, 249). By *Tian* or Heaven, Xunzi is not citing a supernatural force (he is explicitly anti-supernaturalist).[13] Rather, he means that our nature is a product of the natural workings of the cosmos, and it is thereby outside of our control. Interestingly, Xunzi is not saying human nature is neutral and that one may become good or bad while remaining in line with human nature. This is the position of Gaozi, who Mengzi distinguishes his own view from.[14] For Gaozi, human nature may be made good or bad by what happens to it. For Mengzi, human nature is inherently good and can only be thwarted by something external to it. For Xunzi, human nature is chaotic and detestable, and it is only through external means that we can become good.

For all of this, Xunzi is not saying that humans cannot become good. In fact, most of his philosophical energy goes towards explaining how such a transformation is possible. Nature, for Xunzi, only gives us the raw materials with which we must work; it does not set the values, the goals, or even the idea of the good that we should follow. The accomplishments of Heaven alone mean we find ourselves as a chaotic bundle of desires, but we are not limited by the potentials found in our nature alone. Xunzi tells us that "Blue dye derives from the indigo plant; and yet [the dye] is bluer than the plant. Ice comes from water, and yet it is colder than water. Through steaming and bending, you can make wood as straight as an ink-line into a wheel," (Xunzi 2016, 1). In short, a thing's (or a person's) nature does not circumscribe its potentialities, especially when it encounters other things. If the straight wood can become a wheel, then it is possible for the chaotic and detestable human to become good. And for Xunzi, it is the same general process that makes such transformation possible: artifice. To make a wheel from straight wood, one requires the artifice of steaming and bending. To make a morally exemplary human, one needs the artifice of ritual and discourse. In this, Xunzi is a rare thinkers who does not succumb to the repression of technics and instead outlines a way, a *dao* beyond it.

Since one of the raw materials of a human being is desire, Xunzi councils us to preferentially reward or deny desires in order to cultivate them, and to do so according to ritual provisions that maintain society. The key here is cultivation, not merely control. Cultivation takes place through the satisfying of desires in ritually appropriate ways so as to shape future desires. It is not merely giving food to the hungry but giving good food and fostering their enjoyment of it in the comfort of others.[15] This shapes desire so that one no longer just desires the base necessities. Instead, desire is cultivated such that it is now "able to be satisfied in cultured forms through

---

[13] The natural vs supernatural distinction is anachronistic when projected backwards, but Xunzi's various positions accord well with some forms of contemporary metaphysical naturalism. See Xunzi 2016, Ch.17, "Discourse on Heaven" where he argues against omens, divination, and other forms of the supernatural, and in favor of a predictable, non-providential, impersonal regularity to the universe.

[14] See Mengzi 6A1–2. Gaozi's writings, if any existed, are lost.

[15] For an indepth account of the importance of the communal aspect of cultivating desire in Xunzi, see Tang 2012.

ritual,"(Tang 2012, 458). It is actions such as these that, consistently applied over time, shape and transform human beings, (Chong 2003, 223).

Since another raw human material is cognition, we must exercise a proper, though creative and situation-responsive, use of discourse. That is, we must maintain and create useful "names" or concepts that make the right distinctions that allow us to recognize when and how to employ the proper rituals. The holistic synthesizing of human nature, ritual, discourse, and the natural and social environment into a way of life that consistently produces human cultivation and flourishing is Xunzi's version of the *Dao*. This synthesizing does not happen by or according to nature, but only through human artifice and effort (*wei*) (Xunzi 2016, 248–9). However, because this process must be responsive to the materials it has to work with, it is equally not a mere externalization of human will or intention. Xunzi goes so far as to treat artifice as external to the humans it works on, as opposed to their original internal desires. This obviously conflicts with Mengzi's view that good is already internal to us as a potential. For Xunzi, any possibility of becoming good comes from the outside. Thereby, we are not good, nor made good by nature; we only become good through artifice. We have a nature, but that nature is not normative, that is, it does not prescribe or make possible on its own what is good for humans. Instead, to become good, we must change our nature through artifice.

Artifice is then a source of autonomy in a strong sense. Because artifice is external to us, this is an autonomy that is conditioned on our heteronomy. When we create new objects, practices, institutions, etc., we create new ways of shaping ourselves. In such shaping, we transform our nature and gain new abilities (Chong 2003, 220–1). "The sage transforms [their] nature and establishes deliberate effort [*wei*]," (Xunzi 2016, 251). One would only need their tongue slightly in cheek to call Xunzi the first transhumanist.[16] I will defend a modified version of this account below. As it differs radically from classical notions of autonomy (as discussed in Chap. 4), I will generally talk of increasing our autonomy in this sense as "becoming active." The possibility of playing an active role in both applying and creating the artifice that shapes us is the source of any human autonomy worth having, of becoming active in our processes of individuation (in particular, our subjectivation).

5.3 Argument for Xunzi and the Continental Tradition

With this debate in mind, we can return to Garrison's support of Mengzi. As said above, Garrison supports Mengzi over Xunzi for an account of subjectivation based on 1) Mengzi's optimism versus Xunzi's harshness, and 2) the fact that Mengzi does

[16] Note also Xunzi's resonance with the *Xenofeminist Manifesto*, whose last line is "if nature is unjust, change nature."

not maintain Xunzi's problematic internal/external distinction.[17] To begin with, I will grant Garrison that Mengzi's specific political prescriptions line up much better than Xunzi's with the common political values of the continental tradition. For example, Mengzi tells us that to stop banditry, we should be generous with social spending rather than increase state violence to deter theft, (Mengzi 2008, 1A7). Or when a warlord asks Mengzi whether killing a leader is justified, Mengzi says no, but in the sense that if one acts improperly in a position of power, they do not deserve that title, and so if such a ruler's subject kill them, it was not a ruler who was slain, but a criminal, (Mengzi 2008, 1B8). Xunzi, on the other hand, regularly uses particularly harsh metaphors of forcing wood out of its shape, and he is also more likely to council obedience to authoritarian rule.[18] At his worst, Xunzi can read like a prison warden who would think *Discipline and Punish* was a manual of best practices.[19]

Yet, this commonality between Mengzi and continental leanings on political policy masks a deeper conflict between the two views, a conflict not found between most continental thinkers and Xunzi. This conflict happens at the level of human nature. Mengzi's account of the sprouts of goodness is rife with potential abuses of the kind Foucault so adroitly diagnoses. In his view of human nature as inherently good-in-potential, Mengzi naturalizes normativity. This is particularly pernicious given he also maintains the Confucian centrality of ritual, and the traditionalist strain thereof. For Mengzi, the proper rituals are the (even in his time) centuries old ritual order of the Zhou Empire at its height. This order was supposedly created by its highest moral exemplars, the sage-kings. The connection between traditional ritual and human nature is exactly the sort of naturalizing of historical norms of which continental thought, especially its Foucauldian strand, should be extremely

[17] It must be said that I will be focusing only on subjectivation related to Xunzi in this chapter whereas Garrison's work deals with both subjectivation and subjectality. This leads Garrison to a further argument in support of Mengzi: Mengzi's account of the goodness of human nature also implies that we as a species are good at producing societies. That is, we are capable of producing cultural artifacts, practices, and institutions, in short, ritual, that all sediment into society over generations. This sedimentation happens at the level of subjectality. Subjectality and social individuation is the concern of the next chapter. Here we focus on the short-term process of subjectivation. My argument then is not a rejection of Mengzi outright, but rather the claim that for an optimistic understanding of subjectivation, Xunzi is more useful.

[18] There is some argument here as Xunzi appears to tailor his writing to his expected audience. His more authoritarian leanings may be aimed at a general audience so as to maintain order, or he may be writing such passages to rulers to gain their favor. Yet, he also will chastise rulers for not being properly benevolent or relying solely on power; here he is perhaps writing to those rulers, or to other scholars. It seems unlikely Xunzi would intend the latter passages for a more general audience. Even if not blatantly authoritarian, Xunzi is still irredeemable paternalistic.

[19] For example, he councils that "The people can be easily unified by means of the Way [*Dao*], but one should not try to share one's reasons with them. Hence, the enlightened lord controls them with his power, guides them with the Way, moves them with his orders, arrays them with his judgements, and restrains them with his punishments," (Xunzi 2016, 240). Keep in mind that the Way is made up of rituals and techniques to be undergone by and applied to the people. However, this must be softened by his counsel that one cannot force another into moral cultivation or sage-hood.

critical. More generally, on a view such as Mengzi's, knowing human nature would imply knowing what is good for humans. Continental thinkers generally critique such views because they ultimately license an easily abusable relationship between claims to knowledge and exercises of power over others in the guise of knowing what must be good for them. This clearly makes Mengzi a difficult fit for the continental tradition.

One might wonder if given the centrality of the debate over human nature, the maintenance of ritual practice, and an idolization of previous sage-kings as virtuous paragons and the givers of ritual, might the Confucian tradition in general fall to this critique? Yet, such a critique misses the mark if applied to Xunzi. While Xunzi is conservative about ritual practice and the reverence of the sage-kings, he does not naturalize either. Because of this, he already shares important commitments with continental thought that Mengzi does not; namely, that no amount of knowledge of human nature would ever produce a transhistorical set of norms for human good. Ritual, on Xunzi's account is historically situated. He does share the Confucian belief that rituals were the creation of morally perfected sage-kings; however, this does not on his account make such rituals the ideal or only possible good social arrangement. Instead, they are merely the ones that have proven themselves most useful of those that have been tried. The same is true for "names" or concepts. Xunzi recognizes that the sign by which we refer to an object is arbitrary, yet there is an art to crafting the proper names for things because of the content and logical relations between names.[20] Here, Xunzi goes so far to say that if a new sage-king appeared, they "would surely follow the old names in some cases and create new names in other cases," (Xunzi 2016, 237). While this act of creation is conservative in the sense that it requires a moral exemplar and king, it further cements Xunzi as a thinker of the New in that creativity beyond prior rules is required to respond well to the world. It is also worth noting that for Xunzi, and for the ancient Confucians generally, anyone in principle can become a sage, even if sages are rare in actuality (compare this to Aristotle who argues one must be born male, into relatively comfortable circumstances, and should probably be good looking to have a chance at becoming a moral exemplar). While Mengzi and Xunzi agree anyone in principle can become a sage, it is for different reasons. For Mengzi, it is because all humans have the four sprouts. For Xunzi, it is because all humans are capable of transforming their nature (Chong 2003, 215–6).

Despite his reputation for harshness, Xunzi does not think one can be forced into becoming a sage. "They can do it [become a virtuous person or sage],[21] but they cannot be made to do it," (Xunzi 2016, 254–5). In other words, nature and environment alone do not make an ethical exemplar; one must become active in their own subjectivation. Similarly, there are practical counter-examples to Mengzi found in our contemporary world. If our nature contains the four sprouts of virtue, and these

---

[20] The Warring States era thinkers that were the most focused on questions of logic and language were grouped together by Han compilers as "The School of Names."

[21] *Junzi*, or virtuous person (often translated as "gentleman"), and sage are two different stages of moral cultivation, the latter being the highest stage and exceedingly rare.

sprouts depend first on material resources to grow, then one should expect that our contemporary times of overproduction and abundance would give a head start on virtue (in proportion to production and distribution of goods meeting basic needs). While proper ritual would be needed to complete one's moral cultivation, Mengzi's account would lead us to predict that the American middle class, for example, would be moral prodigies merely awaiting proper ritual instruction. This prediction is of course directly at odds with most forms of ideology critique and the forms of subjectivation undergone due to capitalism diagnosed by the continental tradition. While of course not having a critique of capitalism, Xunzi's account of human nature can at least handle this clam because, as Tang points out, "even if we have unlimited resources, we could not live a good life if our nature is not transformed," (Tang 2012, 459). Possibly, contemporary exploitative social relations are harmful enough even in areas of material abundance that they work against the cultivation of the sprouts to the same degree as the animals grazing on Ox Mountain. This is certainly believable to anyone familiar with American suburbs, and so perhaps the issue is a wash.

Finally, we should ask what would it mean for a view of becoming active in our subjectivation to be optimistic? In his debate with Mengzi, Xunzi is regularly seen as the more pessimistic of the two because of his view that human nature is disordered and detestable. Mengzi, on the other hand, tells us human nature has in it the potential sprouts of goodness, and so we only require the proper environment to flower into truly *ren* people. I think this characterization is a mistake. Believing that a potentiality could become actualized is not particularly optimistic. If anything, it is damning to think that something can only become what it by nature already has the potential to become. If a human lacked the four sprouts, if one saw a child fall into a well and did not feel the twinge of concern, Mengzi would be forced to conclude such a person is an aberration or inhuman. They would thereby lack any possibility of becoming good. If optimism is to believe that the world can become better, not just realize its already inherent potential goodness, then Mengzi's view is insufficiently optimistic. Xunzi, on the other hand, gives us a way for something to change or grow beyond its nature, to become what it is not. To believe that what is not good and also lacks the capacity to make itself good on its own can still yet become so is the only form of optimism worthy of the name (whether this optimism is naïve depends on whether such change is possible; this is defended below). Because Xunzi's account argues for this possibility, I agree with Hagen that of the two great followers of Kongzi, Xunzi is the optimist.[22]

Given all of this, I think we have good reason to reject Garrison's first point in favor of Mengzi. To respond to Garrison's critique of the internal/external distinction in Xunzi, we must do more work. This critique points to Xunzi's separation between humans and *tian*, that is, heaven. Again, heaven should not be understood as a supernatural aspect of the world, nor as transcendent to the world. Rather, *tian*

[22] As far as I know, Hagen first makes this claim in Hagen 2003, 90, but the claim reappears throughout his work. See Hagen 2007.

means something like the proper orderly aspect of nature and its workings (even if nature sometimes produces unruly things, like humans). One could make a too-quick analogy between Xunzi's *tian* and Spinoza's God = Nature to get the rough shape of the concept. On Xunzi's account, human disorder means that while their nature is produced by *tian*, humans are by that very nature ill-suited to actually living in accord with *tian*. To produce such an accord, humans create artifice, that is, something external to themselves and the workings of nature. We then find in Xunzi a supposed dualism between nature and artifice, between internal and external. It is here Garrison presses his critique that not only does this dualism present conceptual problems (how is artifice outside of nature?) but perhaps more importantly, it means the "the bodily dimension of human nature is devalued, degraded, despised, detested, or otherwise diminished," (Garrison 2021, 175–6). Garrison has continued this trajectory in his follow-up work, where he turns towards Buddhism to further reject distinctions between internal and external, as well as any lingering notion of the self beyond subjectivation and subjectality, (Garrison 2023). This also amounts to rejecting the perfectionist view of well-being in the Confucian tradition. Insofar as we are searching for the means of opposing the exhaustion of subjectivity and of living well in the age of control, this turn to Buddhism can only be seen as a mistaken detour. Rejecting the separation produced by psychic individuation (as the general form of subjectivation) means falling into the error of reductionist views of the subject and of giving up the goods of heteronomy that control threatens. However, Garrison's challenge that Xunzi does not defend the internal/external distinction, and its potential implications, must be met.

As far as I can see, there is not an explicit response in Xunzi to this problem. There is, however, a space to supplement and work out a response in connection with Deleuze' metaphysics. Simply put, Xunzi's ethical system only works if the new is possible. This is at the heart of his embrace of the internal/external distinction: if we can become good in the way described, it must mean good is not already internal to us. A potential goodness giving rise to actual goodness as in Mengzi is not an example of true becoming. And so, for Xunzi, the source of this becoming must be external to us (here, artifice). Yet, if the good already exists external to us and is simply imprinted onto the one who undergoes ritual while putting in the right effort, then this is not a true becoming either. This is recognized in Xunzi scholarship as the problem of the origin of goodness (Li 2011). Li gives a response to this problem by which the power to institute ritual (or artifice) to carry out non-moral motivations or satisfy non-moral desires may yet result in a good outcome, (Li 2011). Furthermore, undertaking this process repeatedly will cultivate those who institute and follow the ritual. This will then produce a good individual without any innate goodness in them. This solution on its own, however, does not overcome Garrison's objection based on the internal and the external, and it raises the further question of how the new in general is possible. To both respond to Garrison's critique and to fill out Xunzi's account of the *Dao* as an optimistic view of becoming active in our subjectivation, I will turn to Deleuze to provide Xunzi with a metaphysical account of how the new is possible.

5.4 Deleuze and the New

While Xunzi provides some anthropological and ethical material for an account of human improvement, Deleuze can provide the metaphysics to undergird and supplement this account. What Xunzi does not explain is how it is that something can change its nature using means completely unlike itself. Whatever good is, it is not within our nature already, nor is ritual in any way "like the good." We saw that if we simply call the external means, artifice, good, then we similarly fail to account for becoming, which Xunzi clearly wants to account for (as do we). If so, then Xunzi requires a metaphysics where something new can be produced from components that are unlike what is produced. In short, his account requires a non-representational metaphysics of the new. This is precisely what Deleuze has to offer. In addition to this metaphysical supplement, there are also good reasons to think Deleuze and Xunzi would pair particularly well. Both thinkers give desire a fundamental role in their account of human action and subject formation, both thinkers are explicitly against any form of supernaturalist explanation, and both are fundamentally concerned with the formation and transformation of subjects through technology and social institutions.[23]

To respond to this challenge, we must return to the process of individuation, and we must also engage more fully with Deleuze's ontology. For Deleuze, anything at all has three aspects: the actual, the intensive, and the virtual. This is true of whatever may be called an object in the most general sense, including humans, natural phenomena, artificial things, ideas, fictional characters, etc. The actual aspect is the qualities and extensions that it is currently manifesting; it is whatever the object is right now. The virtual is the structural aspect of an object that relates and maintains the various parts of an object.[24] In doing so, the virtual defines a field of potentiality or capacities and affects: the different ways an object can be or become while still remaining the thing it is, the ways it can be affected by other objects, and its capacities to affect others. Deleuze will often talk of these capacities as a field or as defining a phase space for an object. When these virtual capacities are enacted, they alter the actualities of the object. This is what Deleuze calls actualization. Finally, the intensive is the orders of differences that make up these capacities, that is, the ways in which an object differs from itself. Individuation, as opposed to actualization, begins with intensity.

[23] The early Confucians were radically anti-supernatural, especially Xunzi. While "supernatural" is an anachronistic category insofar as the ancient world, especially ancient China, does not carve the world up into natural and supernatural realms, it is meaningful here in that Xunzi rejects the existence of ghosts, the efficacy of divination practices, and so on in a way continuous with many contemporary forms of metaphysical naturalism.

[24] "Virtual" in this sense has nothing specifically to do with virtual reality or the digital objects that we have been discussing, instead coming from a Latin root meaning "power." If digital objects are indeed objects or individuals in the sense we have been using, they will have a virtual aspect in this sense, but this is no different from penguins, humans, the statue of David, the Democratic Party, etc.

What is central here to understanding how we may change what we are and produce new ways of living together is not just the structure of objects, but more importantly, the process of generating new objects. This requires connecting subjectivation to our broader account of individuation. First, let us distinguish actualization and individuation (Deleuze 1994, 247).[25] Actualization is the manifestation of a virtual capacity. In doing so, actualization does not, on its own, produce anything ontologically new. Some virtual event may be actualized for the first time, and this would be remarkable in its own right. However, as the potential for that manifestation already existed virtually, nothing ontologically new has been brought into being. Here, we can rephrase our criticism of Mengzi: in his account, becoming good is merely actualizing an already existing part of our virtual structure. If so, then becoming virtuous would be a mere change in actuality alone, not in the structure of the person. If this was all that was possible, then the best we could do is weight phase space such that one is more likely to approach actualizing the good already within them. While this is good to do in organizing one's life, it is insufficient.

Individuation, on the other hand, produces an ontological novelty: something which has not existed before, nor did its potential pre-exist it in or as a virtual capacity. Insofar as Xunzi is giving an account of how humans can become virtuous, his account is one of individuation, not actualization. We see this in Li's account that actualizing our capacity for instituting ritual is morally neutral; the capacity itself is not good, nor is its actualization. It is rather the way ritual comes to shape those who institute and follow it that good is produced, and this shaping is neither a capacity of the person following the ritual, nor a capacity of the ritual itself. Goodness or virtue is instead a result of an interaction that is found in neither element of the interaction. Therefore, it is a result of individuation.

Drawing from Chap. 2, let us briefly account for how individuation is the process of producing the new. First, we said above that actualization does not produce something ontologically new because the potential for that actualization already exists virtually. If all things that could exist already exist in potential, then whatever could be is already prescribed and de-limited. All possible objects would already exist in a sort of potential pre-formation. This would be the exact confusion of the virtual and the possible that Deleuze critiques, (Deleuze 1994, 211–2), and it would also undermine individuation as onto-genesis. What this means for us here is that individuation cannot begin with the virtual. Similarly, it is for obvious reasons that the process cannot begin with the actual. Recall from Chap. 2 that the first stage of individuation as a process is the formation of pre-individual milieu includes differences of potential energy and tensions, that is, differential relations of intensities. Two or more intensities encounter each other through their differences. Different

[25] While Deleuze on the virtual and actual is in some ways similar to Aristotle on potentiality and actuality, there are fundamental differences. These include 1: Deleuze prioritizes the virtual over the actual in explanations, whereas Aristotle argues the actual explains potentiality, and 2: Aristotle's modality is connected to a hylomorphic account of onto-genesis and Deleuze instead gives a complex account of the interaction between individuation and actualization. Taken together, this accounts for why Aristotle's cosmos is fundamentally fixed whereas Deleuze's is creative.

temperatures on the surface of a material, differences in electric potential between elements in a milieu, and so on. In the right sort of encounter, these differences form gradients, zones of distinctions, etc., all of which amount to forming a field of intensities that coheres.

In cohering, this field generates a "path in advance but in reverse," (Deleuze 1994, 119). Though Deleuze gives the more poetic example of a lightning strike forming from a electric gradient between the sky and earth, this obscure phrase is perhaps better clarified if we think of electrical circuits. Different paths in a circuit will often have different resistances. The current that passes through each branch does so relative to the resistance in that branch. This is not, as popular imagination might have it, because the electrons that make up the current find the right distribution of current through trial and error, either passing through or bouncing off the resistance of one path and heading down other. Instead, the field that organizes the circuit guides the different currents down their respective paths. That is, in the formation and powering of the circuit, the field is quickly cohering itself as it is also guiding the charges that move in the field. The path is formed by the bringing together of the components. Such a path is not formed by the intensive alone. Rather, the intensive encounter sets off a cascade of actualizations. This cascade establishes a continuity of parts through the flow of that which follows the path. It is important to remark that this cascade is not an actualization of the virtual structure identified with a particular object, but rather myriad actualizations of the various components involved outside of pre-existing virtual structures.

There is no virtual whole that encompasses the cascade, it is instead the reverse. This cascade is the second stage of individuation: the formation of a new virtuality. If individuation is completed, it is because the new object coheres into something capable of maintaining itself in at least some limited sense. This results from the production of a structure, that is, of a new virtuality that maintains the components that have been brought together as a whole (Deleuze and Guattari 1987, 256–7). This virtuality amounts to new capacities, including the capacity to resist some attempts to separate the components. In our example, one should expect the circuit, once established, to provide some new potentialities or capacities for acting in its environment. Finally, this new virtuality makes possible new actualizations, and so new actualities are at least potentially brought into being through individuation. Thereby, through individuation, something new is produced. Importantly, this new thing has new capacities that were produced through this process; they were not found previously in the components nor the environment in which the process of individuation took place. To use another example from Xunzi, once the virtuous person is produced, they have capacities they did not have before, such as the capacity to discern what actions would be proper and what actions would not. Such a capacity does not exist in ritual practice or the artifacts involved, but nonetheless ritual practice produces it in the virtuous person. We can even isolate this difference in kind of the new capacity from its components. What the virtuous person gains, in addition to the specific capacities different rituals may produce, is discretion. Discretion is the ability to apply rules or suspend them well by attending to specific features of a situation. The defense of the role of discretion separates the Confucians from the legalists, the latter argued for the strict application of law regardless of situational features.[26] The Confucian *junzi* on the other hand must know when to suspend a ritual prescription, when to make exceptions. We may equate it with *phronesis* in Aristotle's account. The unvirtuous person does not have this ability, and ritual is a series of rule-bound practices. Ritual may teach discretion, especially through rituals of rule breaking where an explicit rule is given but a stronger implicit social script requires we bend, skirt, or blatantly disobey the rule. But by its nature, discretion cannot be found in ritual itself. Only in the interaction, in the individuation of the virtuous person, is discretion produced as a reliable capacity within their virtual structure. This also separates ritual cultivation from disciplinary power in its strictest sense: the internalization of rules is not the end but one set of means among others for cultivating capacities of discretion or *phronesis*.

The particular, local features of an individual human's process of being shaped by their environment, social situation, artifacts, and practices (as opposed to the general structure of biological and psychical individuation as such) is what we refer to as subjectivation. With this account of subjectivation in hand, we begin to see that Deleuze's metaphysics is practically cut to fit for Xunzi's optimism about our ability to change our nature for the better. The new is possible, and it is possible not through some otherworldly power or creator. It is possible immanently through communication and interaction via difference. The conditions of the possibility of different ways of life, different subjects, and different worlds all exist right here, lurking between and within the objects we already have and are. Furthermore, Deleuze's ontology provides a weakened account of the internal/external distinction that Garrison criticizes. If we were to completely reject any internal/external distinction, we would be unable to make the connection to Xunzi. Yet, we can supplement such a distinction to give a reading that it is no longer a metaphysical dualism. To say that virtue or goodness are not internal to human nature is merely to say that whatever capacities one has by birth, virtue and goodness are not among them. However, through subjectivation within social milieus and other forces, a person's capacities, their virtual structure, are changed and shaped by that milieu. We may call that shaping external because it happens through capacities that are not those of the individual. Yet, it is not a hard dualism of the internal/external because each can alter the other, and the boundary is porous. This may indeed be more in line with Xunzi's position than Garrison's dualist reading in that Xunzi argues that ethical transformation is possible. If so, then it seems that such external capacities must have internal effects. One may here make a comparison with Xunzi's student, Han Feizi, who believes it is possible only to manage the unruliness of human nature, not to transform it. Han Feizi may maintain a strict dualism between the internal and the external precisely because he claims that the external cannot produce real change in the internal. In this, he is an example of a Legalist who denies discretion. If instead one argues as Xunzi does that such a transformation of the internal by the external

[26] See Lam 2025 for a contemporary defense of discretion against Legalism that does not depend on Confucian principles.

(and vice versa) is possible, then a strict dualism between the two would be untenable. This version of the internal/external distinction provides support for the goods of heteronomy argued for in the previous chapter, as well as giving us some insight into how they are formed, can be fostered, and may be protected.

In short, we have established a metaphysics that justifies Xunzi's defense of artifice as the means to improve ourselves and our societies. If one can be a wise optimist, it is precisely because of (and given its distinct limits by) the process of individuation. To furnish an optimistic account of becoming active in our subjectivation, we must employ this capacity of the new in some way that is relevant for human projects.

5.5 Building a Way

Where the Deleuzian influence differs from classical virtue ethics is that traditional virtue ethics overvalues the unity of character and character itself. What I mean by the latter is that virtue ethics generally focuses on the product of self-cultivation, a final, resultant virtuous character. This is to value the product over the process. However, we know two things that make this foolish: 1) that a human carries its individuation and subjectivation processes with it as long as it is living. We are never finished products, nor are we ever merely products. We are as much our processes and the relation between those processes and their products,[27] and 2) that we can undergo a radical transformation to who we are that is outside of our control (see Ardoline and Cogburn, Forthcoming). If so, then our character is always vulnerable to change we do not want, that opposes even the best laid plans of self-cultivation. Tying our hopes for a good life to character alone is bound to disappoint all but the luckiest of us. This is the advantage of thinking of character as a sub-concept of a BwO. Character is one more part among other parts, it just happens to play the role of recording, of being assigned responsibility (whether it caused anything or not), and of organizing; these are important roles to be sure, but they are not all we are. Furthermore, the role of contingency and chaos in individuation is one reason to reject the ancient equation between order and the good. Similarly, if *techne* is indeed that which banishes *tyche*, then we must be careful about how we offload processes onto artifacts else we weaken our connection to chaos. Control risks undermining this necessary disorder. What we must turn to, then, is the question of how to mediate our connection to chaos well.

Deleuze famously makes use of the indefinite singular to denote haecceities, particular this-nesses. For example, he speaks of *a* life as a living, impersonal singularity (Deleuze 2005, 27). *A* life is not the human in general, and it is also not the particular person or personality. To explain this, Deleuze gives an example from

[27] This is not to say that we are equally mutable throughout our lives. Childhood is a time of exceptional plasticity whereas adulthood is generally one of greater specialization.

Dickens' last novel, *Our Mutual Friend.* A scene in the story revolves around attempting to save a particularly unlikeable man. The characters act not out of a love of this particular man, nor out of a duty for humankind in general. They are not trying to save any life; they are attempting to save *a* life, this one here, in front of them. They are protecting and attempting to maintain the individuation process itself more so than the specific man it has so far produced. As the nearly drowned man comes back to consciousness, back to his particular personality, he returns to being an unlikeable jerk. No one takes much comfort in having brought this particular man back from the brink. Yet, they have saved *a* life. Their shift from valiant saviors to frustrated workers tracks the stages of the drowned man from *a* life back to being subjectivated as this particular jerk. This is because, on Deleuze's account, the particular covers over the singular (Deleuze 2005, 28–9). There is still *a* life there, it is just hard to see it past an obnoxious personality. What is important about singularity in this sense is that it (1) is not substitutable for another of its type. If he died on the table, a life would have been lost. It cannot be replaced. (2) It is nonetheless somewhat generic or indistinguishable from others of its type. (3) It is in some way in excess of the individual who carries it (e.g., one hopes that when the unlikeable jerk recovers, perhaps he will become a kinder person). What is encountered here is the singularity of the individuation processes one carries with them, what is always underneath their (meta)stable identities. This encounter with the singularity of *a* life may be a Deleuzian rendering of the Levinasian Face of the other (both are excessive encounters with something generic yet singular that compel us), though this would take us into digressive exegesis.

It is in the sense of *a* life that I want to speak of *a* Way or *a Dao*.[28] While the anglophone reader is most likely to associate the *Dao* with the Daoists, most of the Warring States era philosophers used the term. Each thinker or school who did so disagreed with the others over the specifics of the *Dao*, but each claimed an understanding of or wisdom about the *Dao* distinctive of their school. It is a term like the Good or the Just in that nearly every thinker claims them, but no one agrees on what they are. Distinctive of the *Dao* is that it is both a proper path to follow for a good life, and it requires some accord with non-human, usually natural, forces. Hagen's central contribution is to argue that Xunzi has a unique approach to the *Dao* among the ancients. Xunzi argues that "the mind is the craftsperson and steward of the Way [*Dao*]".[29] That is, for Xunzi, the *Dao* is not discovered; it is constructed. Other ways of organizing life are possible, but the *Dao* of Xunzi's context, he supposes, is that which has been crafted by the ancient sage-kings. It is what has been found to work so far. It is open to modification, and it is entirely the work of human artifice [*wei*],

[28] Hereafter I will leave off italicizing the "a" except in cases of potential ambiguity.

[29] This translation is taken from Hagen 2005, 118. Hutton translates the same passage as "The heart is the craftsman and overseer of the Way," (Xunzi 2016, 241). Xunzi, as is common in his tradition, does not distinguish between the rational and aesthetic/affective parts of cognition, leading the term *xin* to be translated as "heart," "mind," or "heart-mind." Thereby, one should be careful to read Xunzi's claims about deliberate human effort and the mind constructing the way not as a rationalist claim about the power of reason alone.

though not of human intention alone. The woodworker must learn the properties of various types of wood, the workings of different tools, etc. to properly join their materials into an artful creation. And so, the craftsperson of the *Dao* must do the same with the objects, the myriad things, human and otherwise, that are brought together into a coherent social order and ways of life.

There are critiques of Xunzi to make here from the point of view of the continental tradition. For one, Xunzi is thoroughly conservative in his historical view. The *Dao* is justified by the fact that the sage-kings shaped it, and only new sages will be able to properly modify it. If we are critical of the idea that historical forces select the most virtuous to rule and instead believe that who ends up ruling is more likely the result of impersonal forces and grabbing of power (and in ways that often attract particularly corruptible personalities), then such reverence is naïve. Similarly, Xunzi's view of the *Dao* is that it is a holistic ordering of a society and the ways of life brought together within that society. If we take the reverence for the sage-kings and the holism together, the view of becoming active produced by such social forces begins to look quite thin. No doubt that such shaping creates new capacities and new possibilities for subjects to act. However, there is very little space to be had for the active shaping by the subject of the forces that will shape them unless that subject happens to be an emperor or an adviser thereof. This is not a dead-end for our encounter; rather, it opens the question: can the *Dao* be fragmented so as to maintain its equally individual and collective nature while opening up the capacity to shape it more generally?

In this synthesis between Xunzi and Deleuze, we can see the start of an alternative conception that would allow for a more engaged possibility for subjects to become active, to craft their activity in an inherently social way. This is not a completed account, but the contours of something new can be glimpsed. Instead of the *Dao*, we may refer to a *dao*, or a way. Let us bring together the various positions worked out so far to describe how this may be done. All objects have a virtual phase space; that is, things are not just what they currently are but have capacities to be otherwise and to be made otherwise. Those capacities mark out a space of possible arrangements and degrees of freedom. Humans are different only in regard to our degrees of freedom and capacities to shape such spaces, but we are equally shaped by them. In bringing objects together, there is a mutual constraining and an opening up of possible arrangements and degrees of freedom. Things can be brought together in ways that are destructive, constructive, or neutral. In neutral cases, things are more beside each other than together; they do not shape a world. In destructive cases, the arrangement destroys itself and harms or destroys the objects that make it up. In constructive cases, a larger phase space is formed; a consistent world or milieu is created. To create a world, we do not simply place objects next to each other. We cobble together a world through inventing discourse, techniques, names, new objects, rituals, etc., each local and appropriate to what is being brought together. We must also discern what must be excluded, removed, or prevented. When we succeed, produced within the phase space are paths that we may travel that will change but not destroy us.

A way is not a metaphor; these are literal yet virtual paths that we traverse through phase space rather than physical space. Because such paths are virtual, they are changed (actualized) by the very walking of them. In doing so, we are changing, our world is changing, all the objects in it are changing; we are neither static nor chaotic because these mutual changes are the very maintenance of an ordered bit of cosmos that produces our path. We maintain a world by traversing a path within it, along with our fellow travelers; there will be fellow travelers as a *dao* is not individual. In crafting a path, we make something others may walk. Yet, in the singularity (not particularity) of such constructions, they remain unfinished and open to possible connections. One chooses for others, but not all others. A *dao* must also then be prepared for radical changes in the context around it. If this is so, then we cannot maintain the disengaged cynical posture, the purely critical reaction, or the hands-off Heideggerian retreat all common to the continental tradition. Instead, we must be engaged, proactively shaping, and offering to and receiving from others (and I do not only mean human others).

Here we must emphasize the materiality of this process. This is best shown in contrast with another set of thinkers who argued for the co-constitution of human subjectivity and artifacts. The post-phenomenologists, specifically Ihde 1990 and Verbeek 2005, make similar claims about the mutual shaping of ourselves through technology. However, Ihde and Verbeek both reduce artifacts to their role as intentional objects that mediate the world. Verbeek in particular argues that the subject and the technical object co-constitute each other. This may sound like what we have argued for so far, but it differs in two important ways. First, in talking of co-constitution, Verbeek means co-constitution within experience (Verbeek 2005, 117–8). Artifacts certainly play this role, but they also play roles at the level of individuation which are not experienced yet structure the possibilities of experience. A pacemaker for example does not need to appear in (or to withdraw from) experience to condition the possibility of a subject. Without blood flow, subjective experience will not be possible for long, but this is a co-constitution that happens at the level of biological individuation, not within or through experience. Second, post-phenomenology reduces technology to an expression of human intentionality (in the phenomenological sense, not the instrumentalist sense treated in Ch. 7). Kleinherenbrink 2024 critiques this tendency as it makes unthinkable the relations between non-human objects in which we are not directly involved. To reduce artifacts to the expression of human intentionality (again, in the phenomenological sense) means artifacts are understandable only in and circumscribed by their correlation to us. Yet how much of our world, of the background scaffolding that makes our particular experiences and consciousness possible, involve complex chains of artifact-artifact relations between which we rarely if ever intervene? To maintain this correlation as the post-phenomenologists do would be to undercut the very externality of artifacts that makes possible the fact that they can play a transformative role in our subjectivation.

A similar failure is found across contemporary continental philosophy, though it is often given in much vaguer terms than the post-phenomenologists. There is a proper worry about our relation to technology, but the retreat from norms mitigates any real outline of a positive project. It hopes for the new while refusing to embrace its own metaphysical resources that provide a guide for producing the new. The response from this corner is always more thought, more discourse, more critique, or, its most positive, for better norms of rational communication, and always a vague call to openness (this latter especially from the phenomenologists, who restrict thought to experience to such a degree that they can only conceive of responding to a problem as waiting for an answer to be shown to them. Heideggerians always already awaiting absent gods). This is foolish. It pins our hopes on the very things being undermined and weakened. Rather, if Stiegler, Deleuze, and Simondon are right and information technologies, artifacts, institutions, etc. all mediate the mind to modulate what sort of thinking is possible, then ignoring these material conditions of thought in favor of thought itself is doomed to failure. Instead, we have shown that we must actively take up the shaping of the material, social, and technical conditions that shape us, that is, our subjectivation. To complete this account, we must return to this shaping. We have embraced the idea that we are mutable creatures that adapt to environments (physical, social, informational, etc.). This is the part of our nature that control exerts its unique power over us by disrupting. We cannot only become active in opposition to this disruption, but do so through constructing the Ways that we travel.

In this, a *dao* is opposed to what Coeckelbergh (2022, 89) calls the narcisistic suicide of the self. This is not a suicide of physical death, but of destroying the self by understanding and living in a non-relational way. Despite its deep interconnectivity, surveillance capitalism fosters this sort of suicide through its atomization of subjects. To build a *dao* is to bring objects (artifacts, people, social institutions, ideas, etc.) together such that they form an ensemble. An ensemble is not itself an object or a unified system with its own identity, though the virtual aspects of its components must harmonize *qua* a mutual co-shaping of each other into a loose consistency. This consistency nevertheless contains tensions, disparities, energy gradients, in short, differences in intensity. A well-crafted *dao* is an ensemble where paths are potentiated, not fully determined or laid out in advance. A proper *dao* without the subjects it will shape (by harmonizing them with the ensemble) must not have the gradient already fully formed in a path traced in reverse. The ensemble without subjects must be able to potentiate the lightning strike, but not cohere the path for it. That is, the path itself takes shape through a human or humans traversing it. In doing so, they provide the condition for establishing these differences into a path (through their own nature as themselves a collection of processes of difference) by establishing communication between some of the parts of the ensemble.[30] This happens at the level of intensity, and so it makes possible the individuation of the new. By allowing one to be the being through which a gradient flows, a proper *dao* enables one to become the lightning.

There are in ideal form three moments here: the assembling of the ensemble, the structuring of the path, and traversing the path. In concrete reality, all three must happen simultaneously. We do not get to stand outside of time or the flow of our lives to assemble the ensemble we will later traverse. We have looked at the first moment of this ideal form as building a *dao*. Now we turn to the latter two moments of maintaining and traversing.

---

[30] This role of allowing communication through difference is what Deleuze calls the Dark Precursor. See Ardoline 2024.

5.6 Ritual

Building a *dao* is how a path is potentiated; ritual practice is how a path is simultaneously followed and maintained. Unlike the ancient Confucians, I do not have a program of proper rituals that one should follow. We are well past imitating the Zhou dynasty. Nor can, like Aristotle, I offer a list of pre-approved virtues or moral exemplars to follow. In this, I agree that we are facing what Vallor calls the technomoral bootstrapping problem, (Vallor 2024, 163–6). This problem has essentially two components: 1) our current and historical systems of ethics and moral legacies are woefully incapable of prescribing good lives, the right virtues, and proper action in our new technologically complex worlds, and 2) like any bootstrapping situation, we can only build new ethical systems out of the materials on hand. A corollary to 1) that complicates 2) is that our current systems of values are part of what got us into this mess. Vallor makes this argument through analyzing the ways artificial intelligence reflects our values, beliefs, etc. back to us. Algorithmic bias, automated wage theft from gig workers, the unequal submission to surveillance, etc. are not technological inevitabilities, but AI-enhanced expressions of the values of our world. If so, then many of worst harms of AI are not the result of "misalignment" with our values, as some commentators contend,[31] but rather are the very results of operationalizing our values and from training machines on data from worlds built by those values.

We are reshaping the worlds we inhabit and we are remaking the types of beings we are. If so, then no simple anthropology, no classical list of habits will be *a priori* derivable as adequate to our new situations. Sages and other exemplars must first be produced through experimentation and encounter, only then can new rituals be produced and new virtues, models, and measures be understood (Xunzi 2016, 251). We have not caught up to our technological worlds. Most of our rituals, cultures, and our ethics assume a world of scarcity, of contingency, of our weakness in the face of the capacities and disorder of the cosmos. In admittedly unequal measure and unjust distribution, many of these issues are vastly mitigated in ways our traditions' ethical thinkers and exemplars would not have thought possible. The reactionary is the one so afraid of what we can and have become that they would return us to the precarious lives our cultures are suited to rather than change who we are. They would salt fertile fields rather than learning to cultivate them for our comfort. This is a danger inherent in merely preserving old rituals.

[31] In AI ethics, this is known as the alignment problem, where it is asked how to ensure that complex algorithms whose workings are often a black box nevertheless are constrained to align its outputs with our values. I agree with Vallor's diagnoses here, though this does not dissolve the alignment problem. It does, however, push back on abuses of the alignment problem where responsibility resulting from collective human action is attributed to algorithms instead. One could call a predictive policing algorithm that re-expresses the racial biases concretized in real estate redlining practices a "misalignment" only if they maintain a complete idealism about values or gerrymander the notion of values such that it only includes what one takes to be the right values rather than the values that guide how policing and distributing property actually happens.

So far, the metaphysics, or meta-ethics to use a disgraceful word, we have laid out has the possibility of ontological novelty. This makes 2) of the technomoral bootstrapping problem less dire as we are not confined to merely rearranging our ethical materials in new shapes; we can produce something new out of them. However, this comes with a trade-off. That which is ontologically new may be imagined, anticipated, protended, etc., but it cannot with any guarantee be *a priori* derived. It can only result from experiment or chance, and this is a metaphysical, not merely epistemic, limit. The critique of the continental mainstream above is that their hesitation in the face of positive prescriptions means continually siding with chance. While defended as intellectual humility, beyond a certain point it is hard to see this as anything more than negligence. This is especially so because embracing experimentation in ethics and politics equally means accepting intellectual humanity as, again, the result of the experiment cannot be derived ahead of time. Rather, I recommend the embrace of ritual *qua* experimentation. By this, I do not mean an ironic stance that half-distances one from the rituals they embrace. This would only hamstring ritual as it means pulling back from a central feature of it, what below we will describe as "feeling into." Instead, we must not only experiment with rituals but given that the ensembles we are within do not remain static, we must have rituals of experimentation. This may at first sound conservative, but this is so only if one misunderstands ritual. Insofar as proper ritual must guard against automation and route repetition in order to protect the possibility of the new, ritual practice is itself experimental. Becoming-active in our subjectivation cannot mean having a particular design down to the minute details on who or what we want to become and staking our ethical and psychological wellbeing on that. We must cultivate joy in the fact of becoming. We must learn to live as the relation between processes outside our control and metastability. Many rituals are required to maintain the world, but maintenance of the status quo is not all that rituals may be. There are possible rituals of the suspension of identity, hierarchy, and tradition; we must have carnivals, celebrations of toppled monarchies, and festivals in honor of Dionysus, the god dismembered.

We cannot solve the technomoral bootstrapping problem by reason alone. Instead, we must undertake an enormous collective practice of ritual experimentation. If both are so, then the best a theoretical work like this one can do is to lay out how proper ritual functions as becoming-active in our subjectivation. Again, we cannot understand ritual primarily as a mere set of rules and practices to be mindlessly followed. Its aim is to produce capable, competent, flourishing people. Hence the *junzi* is one who knows when to break with ritual prescriptions, how to improvise well in response to the complexity and texture of the cosmos that cannot be fully captured by prescriptions, and the sage knows how to create new rituals to harmonize the world and help others flourish. We have already shown that a path cannot be automated for us without preventing the possibilities of ontological novelty or our becoming active. It is as much the encounters with novelty that cultivate one towards sage-hood as it is the proper following of ritual because these two moments are inseparable.

Our active roles must be protected in several ways, not only through the staving off of automation. Ritual must also be aligned with our desire, it must require our feeling into its performance to be done well, and it must facilitate an internalization or re-embodiment of technical memory. If desire is the basic building block of subjectivity, then the rituals we construct must accord with it to some degree. Desire draws us into the practice and maintains us throughout. However, effective rituals are not desire-satisfaction, but desire modulation and organization. Without ritual, our desires are chaotic and in tension with each other. Furthermore, control disrupts the organization of desire and mutes its force. First, a ritual practice must be compelling; it must relate to something we desire. This is the failure of ascetic and self-punishing moralities in that they deny desire rather than engage it (with the important caveat that one can come to desire asceticism, hence many BDSM practices and their continuity with mystical practices of religious ecstasy).

Once one is drawn into the practice via desire, the practice must connect the initial compelling desire with something else. For example, if a teacher is entertaining enough to draw students into paying attention to them, their job is to then direct students through practices that turn that attention towards the subjects they teach. This is successful when the student has developed a desire to learn more about the subject regardless of the entertainment skills or charisma of any particular teacher. Good ritual does not just stoke desire; it must restrain enthusiasm when it is excessive or destructive. It must also provide meaningful satisfaction. In short, it must invite, modulate, and deepen desire towards various goods. In the modulation of desire and feeling, ritual can establish pretense and oppose authenticity, thereby disrupting some of the confessional impulse on which control feeds. If this discussion seems to take on a religious or exoticizing tone given the connotation that "ritual" has in its colloquial English usage, I note here that the Confucian usage includes things like the best practices for a task, institutional bylaws, and bureaucracy. While in many cases contemporary malaise results from a lack of ritual, it is as frequently the result of us not understanding that we were crafting rituals that we were submitted to or knowing what they should have been for.[32]

This modulation of desire cannot be done solely externally; those undertaking ritual cannot be coerced into it, no can they merely go through the motions mechanically even if they desire the outcome or some particular feature of the ritual practice. Rather, they must "feel into" the ritual, by which I do not mean just that they are "up for it," but also that they affectively invest in its performance. To feel into a ritual is to extend one's emotional capacities into it, to engage with it unironically (unless the ritual calls for irony), to allow oneself to be, within reason, emotionally swept up into it. This has a few important consequences. First, this is part of how one plays a role in the ensemble at the level of intensity, producing gradients of emotional difference. Second, it makes possible affective complexification. Third, this gives

[32] Here I disagree with Han 2020 who describes the cause of many modern woes as the disappearance of ritual from contemporary life. I see this as only part of the problem because I take it that no human societies could exist without ritual in the broad sense. We have rituals that entrain consumers under capitalism, structure the employer-employee relationship, subsume education under "practical" goals, nurture repetitive actions through addicting app design, and so on, and they are profoundly damaging.

ritual a unique relation to power. Note that ritual is a form of power in Foucault's sense; it is conduct that acts on our conduct. However, ritual is power that we are not simply subjected to but subjectivated through, and only insofar as we take up an active role in its performance. In this sense, proper ritual practice has something in common with play; just as anything that one is coerced into is not play, anything which one is coerced through the motions of is not ritual. A well-constructed ritual harmonizes power and desire.

This feeling into the ritual is not only a cognitive difference, but a metaphysical one as well. Feeling into well is a sign that we are intensively coupling with the ensemble we are traversing. Intensive coupling distinguishes a proper ritual from one that is excessively automated. In an automated path, one is simply moved through an ensemble (or a system if it is automated to a high degree of technical individuation). One then does not establish connections through being within the tensions that give rise to structure. Rather one moves through it as if a fully formed individual on a conveyor belt, at best actualizing some of their virtual capacities, but never to a degree that forces them into the intensive, never being provided the energy required to form new structure. In being so moved, one becomes passive, deskilled, disempowered, and ultimately more vulnerable to the world. Unfortunately, such are the design principles of the contemporary workplace.

Wojtowicz and Dedeo 2025 analyze how generative AI erodes trust precisely by automating away our active role in certain rituals. Take their example of an apology. A simple "sorry" will not do for an important apology. One must say or write something that shows they have thought it through, understood the offense, and so on. If done well, they produce something heartfelt. This is a sign of the care taken and of effort promised to not repeat the offense in the future. If generative AI produces text that reads just as well, but one can produce it without any real thought or any real feeling into the ritual, this form of signaling breaks down. One who takes this route will not be subjectivated through this process; they will not be meaningfully changed. Building such a tool into the maintenance of an ensemble then removes the ensemble's need for humans to act as intensive couplings. This has further social implications that threaten the consistency of the ensemble itself. Without the proof of care, or the reasonable expectation of it, one no longer can trust the product to be a sign of the right sort of process. That is, insofar as one automates these rituals of social maintenance, they erode the trust and care that are the very fabric of sociality.

Finally, ritual must involve an engagement with and a re-internalization of something retained in technical memory. Insofar as the transformative power of ritual comes from something external, and any ritual practice is going to be mediated through artifacts, technical memory must play a role in that external transformative force that is then internalized. Homer's *Illiad* and *Odyssey* did not shape western cultures when they were finally written down, but when they were performed and read. In not merely moving through a system as a mechanical part, but in playing a role that allows components of an ensemble to communicate and individuate themselves, one becomes a conduit for technical memory. This can be cognitive, affective, physical, or any other way in which one may engage in an ensemble. A skill may be re-embodied and practiced, knowledge may be transported and transformed, an artistic expression may draw others together, and so on. In cultivating oneself, one must care for collective memory, for artifacts, and for the material supports of our worlds.

Here is a basic example: smart phone app design aims at undermining the interval. They habituate us so that opening an app happens before consciousness takes account of what is going on; it becomes compulsive. This undermines attention and willpower (if such a thing exists). The idealist response here is to either express greater willpower or to think critically about what we are doing. The interval is required to do either of those things. The idealist's advice is therefore worthless. Take instead the example of an app like One Sec, which allows you to list certain apps to then prompt a second window on launch that asks you to take a breath and consider if you actually do want to open that app. This delay and challenge protects the interval in which reflective thought can affect our actions and provide the opportunity to retrain habits. Such an app prompts one to engage in a small digital ritual; it is a concretized aid to self-cultivation. In doing so, we gain a moment of active engagement with our desire, we take a moment to appreciate this engagement, and technical memory helps us re-embody attention. This is an extremely minor example, but insofar as it takes account of the material nature of such interactions, it is worth more than all the pages of idealist theory.

Ritual is often misunderstood as mechanical, lifeless, and lacking in the spontaneity so often prized in continental thought. Proper ritual is none of these things. It is vital, it opposes the reduction to automation, and it cultivates the conditions of spontaneity and the virtues needed to improvise well. At some point in the process of proper ritual practice, there will be a moment where all that was necessary and coercive disappears, where the heteronomous excess of all around you energizes a moment of pure potentiation, a unique bifurcating point in a changing phase space. You will be left in a sovereign moment of indetermination. There you are beyond even will; what is a will but self-coercion? No, in this moment, you merely do what is most in your nature, that is, you reshape your nature. That moment will be both action and reward, your only true taste of freedom.

5.7 Conclusion

This chapter begins the positive aspect of our project through an account of constructing ritual as a powerful form of becoming active in our own subjectivation. Such an account can respond to control because it does not operate mainly at the level of the subject or their will, but involves engagement at the level of the formation of the subject itself. While one is shaped by the institutions, practices, and societies they inhabit, their possibilities of response are not completely circumscribed by such social forces. One's possibilities of response may even be augmented by that structure beyond what it would circumscribe, as ritual practice can produce ontological novelty. This is what we meant by an optimistic account of subjectivation, as opposed to the often defeatist, pessimistic, or quietist accounts common in the continental tradition. What the encounter between Deleuze and Xunzi has shown is that such optimism can be reasonably grounded if we accept that natures can be transformed and that the new is possible. Understanding artifice and individuation, we see that the question of "if subjectivation means we are shaped to our very core by the world we find ourselves in, where can we find the means to a better world?" is a deeply flawed question.

We have also derived some practical advice. This advice, in summary, is that the new is produced in encounters, but that we as humans are rarely transformed quickly. Producing something outside the current status quo is a condition of autonomy through subjectivation, but it is only transformative for us when it is integrated into and thereby alters our ways of life. It is this symbiosis between Xunzi and Deleuze that makes visible a deep lesson often missed by interpreters of the latter thinker: the point of the new is not to chase novelty; it is to use what is new to change the path we are walking and so craft a life worth living.

Ritual is not merely individual. It is the pivot point between the self and society, as much the maintenance of our collective unconscious as it is of ourselves and our artifacts, (Garrison 2023, 243–6). To explore this, we turn now to the question of becoming active in our social individuation.

References

Ardoline, Michael J. 2022. Building a Way: Becoming Active in One's Own Subjectivation through Deleuze and Xunzi. *Philosophies* 7 (5): 98. https://doi.org/10.3390/philosophies7050098.

Ardoline, Michael J. 2024. *Deleuze, Mathematics, Metaphysics: Difference and Necessity*. Edinburgh University Press.

Aristotle. 2019. *Nicomachean Ethics*. Translated by Terence Irwin. Hackett Publishing.

Coeckelbergh, Mark. 2022. *Self-Improvement: Technologies of the Soul in the Age of Artificial Intelligence*. Columbia University Press.

Chong, Kim-Chong. 2003. Xunzi's Systematic Critique of Mencius. *Philosophy East and West* 53 (2): 2.

Deleuze, Gilles. 1994. *Difference and Repetition*. Translated by Paul Patton. Columbia University Press.

Deleuze, Gilles. 2005. *Pure Immanence: Essays on a Life*. 2nd ed. Translated by Anne Boyman. Zone Books.

Deleuze, Gilles, and Felix Guattari. 1987. *A Thousand Plateaus: Capitalism and Schizophrenia, Vol 2*. Translated by Brian Massumi. University of Minnesota Press.

Foucault, Michel. 2015. *About the Beginning of the Hermeneutics of the Self: Lectures at Dartmouth, 1980*. Translated by Graham Burchell. University of Chicago Press.

Fuggle, Sophie. 2013. Stiegler and Foucault: The Politics of Care and Self-Writing. In *Stiegler and Technics*, ed. Christina Howells and Gerald Moore. Edinburgh University Press.

Galison, Peter. 1994. The Ontology of the Enemy: Norbert Wiener and the Cybernetic Vision. *Critical Inquiry* 21 (1): 228–266. https://doi.org/10.1086/448747.

Garrison, James. 2021. *Reconsidering the Life of Power: Ritual, Body, and Art in Critical Theory and Chinese Philosophy*. SUNY Press.

Garrison, James. 2023. Reconsidering the Life of Power: Buddhist Perspectives. In *Buddhism and the Body*, ed. W. Kenneth. Holloway. Brill.

Hagen, Kurtis. 2002. Xunzi's Use of Zhengming: Naming as a Constructive Project. *Asian Philosophy* 12 (1): 1. https://doi.org/10.1080/09552360220142243.

Hagen, Kurtis. 2003. Artifice and Virtue in theXunzi. *Dao* 3 (1): 1. https://doi.org/10.1007/BF02910342.

Hagen, Kurtis. 2005. Sorai and Xunzi on the Construction of the Way. *Asian Philosophy* 15 (2): 2. https://doi.org/10.1080/09552360500123683.

Hagen, Kurtis. 2007. *The Philosophy of Xunzi: A Reconstruction*. Open Court.

Han, Byun-Chul. 2020. *The Disappearance of Rituals: A Topology of the Present*. Polity.

Heidegger, Martin. 2017. 'Only a God Can Save Us': The Speigel Interview (1966). In *Heidegger: The Man and The Thinker*, ed. Thomas Sheehan. Routledge.

Hui, Yuk. 2016. *The Question Concerning Technology in China: An Essay in Cosmotechnics*. Urbanomic.

Hui, Yuk. 2024. *Machine and Sovereignty: For a Planetary Thinking*. Univ Of Minnesota Press.

Ihde, Don. 1990. *Technology and the Lifeworld: From Garden to Earth*. Indiana University Press.

Ivanhoe, Philip J. 1991. A Happy Symmetry: Xunzi's Ethical Thought. *Journal of the American Academy of Religion LIX* 2:2. https://doi.org/10.1093/jaarel/LIX.2.309.

Ivanhoe, Philip J. 2017. *Oneness: East Asian Conceptions of Virtue, Happiness, and How We Are All Connected*. Oxford University Press.

Kleinherenbrink, Arjen. 2024. Getting Real about Artifacts: OOO, Post-Phenomenology, and Technology. In *15 Years of Speculative Realism: 2007–2022*, ed. Charlie Johns and Hilan Bensusan. Zer0 Books.

Li, Chenyang. 2011. Xunzi on the Origin of Goodness: A New Interpretation. *Journal of Chinese Philosophy* 38 (s1): s1. https://doi.org/10.1111/j.1540-6253.2012.01688.x.

Machek, David. 2025. *Self-Cultivation in Chinese and Greco-Roman Philosophy: Nourishing the Heart/Mind and Playing Roles*. Bloomsbury Academic.

MacIntyre, Alasdair. 2007. *After Virtue: A Study in Moral Theory*. 3rd ed. University of Notre Dame Press.

Mengzi. 2008. *Mengzi: With Selections from Traditional Commentaries*. UK ed. Translated by Bryan W. Van Norden. Hackett Publishing.

Monahan, Michael J. 2011. *The Creolizing Subject: Race, Reason, and the Politics of Purity*. Fordham University Press.

Mozi. 2014. *The Book of Master Mo*. Translated by Ian Johnston. Penguin Classics.

Norden, Bryan W. Van. 1992. Mengzi and Xunzi: Two Views of Human Agency. *International Philosophical Quarterly* 32 (2): 2. https://doi.org/ipq199232212.

Ott, Margus. 2024. *Resonances of Neo-Confucianism*. Palgrave Macmillan.

Sealey, Kris F. 2020. *Creolizing the Nation*. Northwestern University Press.

Stiegler, Bernard. 2015. *States of Shock: Stupidity and Knowledge in the 21st Century*. Polity.

Tang, Siufu. 2012. Self and Community in the 'Xunzi'. *Frontiers of Philosophy in China* 7 (3): 3.

Vallor, Shannon. 2024. *The AI Mirror: How to Reclaim Our Humanity in an Age of Machine Thinking*. Oxford University Press.

Varela, Francisco J. 1999. *Ethical Know-How: Action, Wisdom, and Cognition*. 1st ed. Stanford University Press.

Verbeek, Peter-Paul. 2005. *What Things Do: Philosophical Reflections on Technology, Agency, and Design*. 1st ed. Penn State University Press.

Wojtowicz, Zachary, and Simon DeDeo. 2025. Undermining Mental Proof: How AI Can Make Cooperation Harder by Making Thinking Easier. *Proceedings of the Thirty-Ninth AAAI Conference on Artificial Intelligence and Thirty-Seventh Conference on Innovative Applications of Artificial Intelligence and Fifteenth Symposium on Educational Advances in Artificial Intelligence, AAAI'25/IAAI'25/EAAI'25* 39 (August): 1592–1600. https://doi.org/10.1609/aaai.v39i2.32151.

Xunzi. 2016. *Xunzi: The Complete Text*. Reprint edition. Translated by Eric L. Hutton. Princeton University Press.

Yin, Jing. 2013. Becoming-Animal: Becoming-Wolf in 'Wolf Totem'. *Deleuze Studies* 7 (3): 3.

Chapter 6
Group Cultivation Under Control

"Culture has constituted itself as a defense system against technics,"

—Simondon 2017, 1.

It is not enough to have an individual solution because control shows us that we have never been atomic individuals. This chapter addresses this concern by turning to the cultivation of social groups, or what Deleuze and Guattari call in *What is Philosophy?* the summoning of a new people. We may make use of a bit of synchronicity here in that "the Great Learning" (*Daxue*), a short classic Confucian text, declares that the *Dao* lies in "making the people new" (as well as in making virtue shine and resting in the good). The text then lists a series of steps that connects properly ordering states to basic self-cultivation in a series of steps. It concludes that this process of setting the world in order begins with self-cultivation. If we understand this in an overly individualistic way, it is hopelessly optimistic. If, on the other hand, we understand that self-cultivation is a becoming active in and a taking care of the various processes that shape us, then it is plausible insofar as we can intervene in our social individuation as well (even if it remains idealistic to think that there is a proper order for the world). A *dao* is already social, but some account is needed of how the social itself can be reshaped by those within it in a way that does more than merely reproducing the subjects of power. This chapter explores this issue by looking at two other accounts of the formation of peoples, Sylvia Wynter's sociogeny and Li Zehou's subjectality, and the role technology plays in both. Ultimately, it argues in favor of subjectality's combination of materialist and aesthetic aspects of group-cultivation.[1]

[1] The immense damage done to sociality by control and its resultant problems is an enormous topic all its own. Insofar as this book is centered on the exhaustion of subjectivity, our aim in this chapter is limited to explaining the metaphysics of social individuation on individual subjectivity and

M. J. Ardoline, *Interiority Against Control*, Sophia Studies in Cross-cultural Philosophy of Traditions and Cultures 47,
https://doi.org/10.1007/978-3-032-16245-8_6

6.1 New People

As we saw in Chap. 2, psychic and social individuation are in a continual feedback loop. Who we are shapes what we are a part of, which shapes who we are. In fact, at the level of individuation processes, we are not separate or even cleanly distinguishable from our environment, artifacts, and other people. In affecting us at the level of individuation, the techniques of control also make possible forms of power that act on social and technical individuation, providing the ability to predict and intervene in group formation and the flow of the machinic phylum as well. Worse yet, some tech oligarchs realize this. Take for example the far right's embrace of the principle that politics is downstream of culture; to succeed politically, first reshape culture. To do so effectively, exert power over technology *qua* technical memory as the support of culture. Peter Thiel lays out this vision thusly: "We could never win an election on getting certain things because we were in such a small minority, but maybe you could unilaterally change the world without having to constantly convince people and beg people and plead with people who are never going to agree with you through technological means, and this is where I think technology is this incredible alternative to politics," (Thiel 2010).

This technological intervention into groups, cultures, etc. is once again the result of technology as technical memory. This reverberates from the mass to the dividual level through the way technical memory shapes imagination, that is, technical memory as schematism, (Stiegler 2011). Pulling from Kant's account of the transcendental structure of the mind, schema are what the imagination uses to mediate between particular sensations and general concepts. Adorno and Horkheimer lamented that the culture industry usurped certain powers of the mind in that by shaping imagination and memory, the culture industry effectively colonized the imagination, specifically the schema, of the masses, (Horkheimer and Adorno 2002, 98). However, if Stiegler is right, then technology has always been a form of memory and our schema have always been partially the result of technological (and social) mediation. Our experience and knowledge of the world, and our actions downstream of that, are shaped in part by the technical memory we engage with. Earlier, we argued for the central role technical memory plays in group formation. Explicating this fact through its role in the imagination then leads us to the centrality of aesthetics for group cultivation. To do so, let us turn to Deleuze and Guattari's discussion of summoning a new people.

Deleuze and Guattari give little more than cryptic remarks on the idea of summoning a new people in their final work *What is Philosophy?* They claim that when philosophy turns to the task of analyzing capitalism as the milieu in which it finds itself, it turns capitalism "back against itself so as to *summon forth a new earth, a new people*," (Deleuze and Guattari 2009, 99). This is philosophy's untimely nature, and it is in this way that all political philosophy is utopian. Its end goal is not new

identifying how and what it means for one to become active in that process. Collective action and similar topics are fundamentally important here, but outside the scope of this work.

ideas, but ultimately, new peoples, new ways of life, and new ways of being human. "Becoming stranger to oneself, to one's language and nation, is not this the peculiarity of the philosopher and philosophy," (Deleuze and Guattari 2009, 110). Art summons a people as well (and there is a hint in the text that science does so too), (Deleuze and Guattari 2009, 218). Art does so through creative production, freeing some bit of sensation, some percept, some already established form of meaning and understanding, and using it to make something that adds to chaos. As I said, this is all quite cryptic.[2] We can explicate this a bit more through Deleuze and Guattari's work, but from there, I will turn to two of the great thinkers of group cultivation, Sylvia Wynter and Li Zehou, to complete the account.

We must first separate summoning from creating. Creating is taking raw materials and forming them into a finished product, here, humans into social groups. Summoning is instead the cultivation of the conditions for new ways of life and stoking the desire to engage in them. Creating a new people is a horrifying and totalitarian enterprise; we see such attempts and failures throughout history. Summoning, on the other hand, requires that those who respond become active in their role in group formation. They must do it, and they cannot be made to do it. Here we may seem to be tending back to Garrison's reading of Mengzi, where one ought to cultivate an environment in which others flourish, rather than try to act on individuals directly to shape them. There is some truth to this, but we remain within the purview of Xunzi and the new because the new people we aim to summon do not already exist in potential; they are not waiting as Mengzian sprouts. Philosophy, art, and science cannot create a people, they can only "summon" them, or attempt to call them forth, (Deleuze and Guattari 2009, 110). The technomoral bootstrapping problem gives us further reason to embrace summoning over creating. We do not know what sorts of peoples are adequate to living well in the age of control, and we won't know until they come to exist. Creating a people, if it were possible to do well, would still require a blueprint we do not and cannot have. The question then is what form does this summoning take? What can one do to become active in the calling forth of new peoples without falling into coercive fantasies, or worse. That is, how does one engage well in the processes of group cultivation that they are already caught up in?

The biopolitical model for the creation, or rather "recognition" of a people has been nationalism. Under this model, a people is declared based on a tenuous claim to some biological and cultural unity, usually through positing a special relation to a plot of land and a vaguely shared mythologized history (mediated through technical memory). This is justified as the recognition of groups who already existed, but recognition is merely part of the process of generation. Recognition has always been a farce; one need only look at the history of colonialism and see how colonizers separate the colonized, often generating new ethnicities, deputizing some and marking out others for greater exploitation. Embracing nationalism is a clear mistake. Even when group resistance to hegemonic powers takes this route, they form themselves

---

[2] For an excellent in-depth treatment of Deleuze and Guattari on the people to come, see Scholtz 2015.

into the sort of enemy that a dominant power wants. The logic of nationalism always tends towards genocide. Furthermore, whatever Deleuze and Guattari mean by a people, it cannot be in the nationalist sense of the term. A nation, no matter how small or exploited on the world stage, is always "major," claiming to be pure, homogenous, autochthonous, etc. But the people that art and philosophy summon is "not the one that claims to be pure but rather an oppressed, bastard, lower, anarchical, nomadic, and irremediably minor race," (Deleuze and Guattari 2009, 109).

The above are reasons to reject nationalism generally. However, one might wonder if there is something to be gained by the re-assertion of nationalism in the face of control, like the reassessment of discipline. Could it be of strategic use? This would be a grave mistake. By its nature of modulation and the vast expansion of access to mythic histories stored in technical memory, control breaks down the boundaries of territory and connections to place. That is, it undercuts the very conditions of a nation conceived as a stable population tied to a territory that it then has unique political claim to. Despite the damage done in the process, we owe control thanks for this just as we do for its undermining of the supposedly autonomous, atomic individual.

In response to the increasing modulation of populations and deterritorialization of control, many have lamented the loss of rootedness, of that special connection to a place. I'm skeptical of this. I take it that it misidentifies two real goods but misunderstands them under the biopolitical model. The actual goods are having 1) a comfortable way of moving through some portion of the world and 2) a positive background feeling towards arbitrary social connection in these spaces. By the latter, I mean the condition of being well-disposed to many of the people around us with whom we happen to be only arbitrarily connected to, people we did not seek out, and with whom we often do not even know if we have common interests, shared values, or the like.

Rootedness assumes that these goods can only be had through 1) a stable tie to a common territory produced by 2) sharing a common culture. These are fundamental mistakes. First, nomadic forms of a life show 1) is false. Deleuze and Guattari claim that nomads never move because they take their worlds with them (Deleuze and Guattari 1987, 482). What this means is that they have forms of life that are responsive to the places they traverse, that allow them a habitual, communal way of life, yet are not tied to any particular portion of land. For 2), Iris Marion Young's account of the city puts the lie to the idea that to have an existentially satisfied sense of belonging, one must be a part of a cultural homogeneity, (Young 1990). In a city, one most often lives more like a nomad though not because one moves often. Rather it is because the city itself changes underneath one, and because to traverse a city is to move through a patchwork of plural spaces. A city maintains "social differentiation without exclusion," (Young 1990, 238). Similarly, positive arbitrary social connection is nurtured in cities not by commonality, but by an erotics of plural spaces, (Young 1990, 239–40). Not all cities maintain this, such as American cities that continue to suffer under the combined forces of rational planning and legacies of segregation. A city maintains a healthy *eros* when its citizens are regularly drawn beyond their familiar spaces for more than transactional reasons, do so with minimal exploitation, and are rewarded with enjoyment for doing so well. Is it any surprise, then, that reactionary nationalists always come to hate their own great cities and elevate some conception of a rural *volk*? If any of this is understood as a rootedness in a city, it is only a result of how deeply nationalism has infected our political imaginary. Nor is this merely a model for the physical world, as Feenberg 1995, Ch. 5 embraces the city as a model for online spaces. However, the question is not simply how to reproduce the model of the city into other spaces, but what are the forces that nurture these goods regardless of the types of space?

The nurturing of positive arbitrary social connection is understood by the Confucian tradition to happen through first practicing filial piety. By attending to family, one develops their humaneness and extends it to their society. This can be read in the nationalist way if families are seen as the basic unit of shared culture, but this is a mistake. Rather, it is the arbitrariness of family that helps cultivate the extension of humaneness to broader groups. We do not choose our parents, our siblings, cousins, our children, etc. They are often quite unlike us in values, lifestyle, temperament, etc., and we are often better for this. Add to this the fact that when family members are chosen, it is often through connections of chance and desire (who our siblings date or marry, the friends who become a part of the family, etc.), and they can be quite anarchic indeed. This is not to extol families as the only or the best place through which to cultivate these feelings, nor to denigrate the many reasons one might avoid or break off contact with their families completely (Mengzi may disagree with me here), but to note a ubiquitous starting point to do so.[3]

With a sense of what we are aiming to address, we should now turn to the question of how it is that group are formed in the first place and return to Simondon on social individuation. In this sense, we should think of groups as individuals as well, but individuals whose parts are human beings, mediating artifacts, practices, etc. This also gives a general insight into the conditions of group formation if we are to stave off some of the homogeneity that globalization produces. We need energy and a relaxation of competition. The high stakes of global capitalist competition make a small number of problems into the fundamental attractors that nearly all groups must respond to (especially the need to grow profits). Scarcity of available energy (in all its senses) constrains possible solutions to these problems. The commonality of problems, the intensity of the force of those problems, and the greater dispossession from people of the means to respond to those problems in new ways ensures increasing global homogeneity. By analogy, all state armies are effectively organized the same now. The great difference in war today is between state military

[3]The *Mengzi* contains several sections which retell and interpret the stories of sage-king Shun. Shun comes from a humble background, especially for a man who becomes emperor. Unique to his story is his horrible family. For example, when Shun's half-brother mistakenly believes that Shun has died in a fire, he does not mourn but immediately goes to claim Shun's property (Mengzi 2008, 5.A.2). Shun however, is the very model of filial piety, perhaps partially because of how difficult it was to navigate being pious to his family in a ritually appropriate manner (see Mengzi 2008, 7.A.35). There is the implication that learning to extend humanness and ritual propriety in such a dire family situation is part of what makes him a sage-king and worthy of being emperor. I see no reason one should follow Mengzi this far, especially given differences in family structures, their role in society, and the like.

forces and insurgents, but as far as any battlefield operations go, all state military forces have effectively converged on similar organization, tactics, equipment, etc. Why? It works and war is life or death, so the pressure to pursue what works is incredibly high. Fascists love war because in the entropy of total war, all societies become the same.

Here I share Hui's values of technodiversity and epistemic diversity, but not his tying of this diversity to current, broadly nationalist differences, nor his diagnosis of the source of homogeneity. Hui 2024 takes his Heideggerian approach and merges it with Hegelian political philosophy. The driving force behind global homogeneity on such an account is the development of Western metaphysics. This has all the issues for which we critiqued Heidegger's account of technology as *Gestell* in Chap. 2. Instead, I take the source of increasing global homogeneity to be the result of the hegemonic force of the profit motive, especially through the techniques of private equity and ever increasing financialization. It is the drive not just for profit but for ever growing profits that is a constant forcing of the organization of production into ever lower energy states. Homogeneity is high entropy. That is, homogeneity is disorder qua the lack of structural differentiations. This is the lesson of the second law of thermodynamics that is so difficult to learn because it is in conflict with the assumed equations of chaos with diversity and of order with sameness that is common across many cultures.[4] Order is distinction, fending off disorder means establishing order, or maintaining structures of distinction. Disorder is the breakdown of these structures. The lowest energy level of a system is its most homogeneous, its most disordered. Capitalism is the great homogenizer. It accelerates towards absolute deterritorialization. This can be a useful movement as not all distinctions or structures are good (racism, patriarchy, etc.), but complete homogenization is death, whether of a person, group, or world. "Corporate efficiency" is the name for the financialization of Protestant hatred for this world.

What then opposes homogeneity? The ready availability of energy beyond the lowest energy state of a system and the relaxation of competitive pressures. From here, however, Simondon's account is incomplete. Nor do Deleuze and Guattari detail what it is to summon a new people beyond the centrality of art, science, and philosophy. We have the general conditions of social individuation, but little is worked out beyond this, especially regarding how particular circumstances shape the process. For this, we can turn to the two great thinkers of social individuation: Sylvia Wynter and Li Zehou.

6.2 Sociogeny

In a project similar to what we have seen already with Simondon and Stiegler on social individuation, Wynter gives an account of subjective (or rather intersubjective) structures whose genesis cannot be attributed to biology, individual

[4] For an excellent extended meditation on this, see Mussett 2022.

psychology, or the combination of the two. Her account is incredibly detailed, engaging with anthropology, evolutionary biology, cognitive science, and analytic philosophy of mind in addition to phenomenology, post-structuralism, literary studies, and the like. Given all this, my reconstruction and analysis here will be unfortunately cursory and partial. In seeking the metaphysics of social individuation so that we may have an account of how we can become active in the process, I will make use of a particular virtue of Wynter's account of sociogeny. In developing her account, Wynter most completely connects this process of group formation to its subjective manifestations. We can make use of this to trace our way back to this process's conditions.

Wynter's account picks up a project began by Fanon, who argues that human experience is not only the result of ontogeny and phylogeny, but also of sociogeny. In short, we live in societies, and we must produce societies to live, (Wynter 2006, 134). "The sociogenetic principle, put simply, claims that we are the kinds of things that we are… as a result not only of the biological processes that produce us, but also as a result of the social processes and conditions in which that biology functions and through which we are produced," (Monahan 2024, 264). Sociogeny is the possibility of producing subjective commonalities (similar forms of inner life) but also subjugated or subjugating minds. It sets soft boundaries on what we are likely to think, feel, and experience. Here Wynter aligns with our wider project as this process is possible due to certain technology, or rather, through understanding culture as itself a form of technology. "We can *experience ourselves as human* only through the mediation of the processes of socialization effected by the invented *tekhne* or cultural technology to which we give the name *culture*," (Wynter 2001, 53, original emphasis).

The result of sociogeny is the production of a certain "genre" (in Wynter's terminology) or style and way of being human that is unique to the group produced. This is both phenomenologically verifiable, yet traceable to external, technological, political, and symbolic forces. Take for example the chapter from Fanon's *Black Skins/White Masks* on "The Lived Experience of Blackness." This chapter analyzes what it is to be a subject racialized as black under colonial conditions. This is the analysis of a particular way of being human. Insofar as it is not universal to all humans, it cannot be the result of phylogeny. Yet insofar as it is intersubjectively verifiable or deniable by others of the same genre, it is not a mere individual report or result of a particular ontogeny.

Riffing on Nagel, Wynter argues that sociogeny as a third process is required to explain how there is any such phenomenon as "what it's like to be black," (Wynter 2001). "The processes by which we produce our societies in order to live are the same auto-instituting processes by which we at the same time produce ourselves as this or that modality of an always already socialized, and therefore sociogenic, kind/genre of being human," (Wynter 2006, 134). In other words, the production of a genre of human is, among other things, the production of unique phenomenological ways of being. Sociogeny is not a determinate process, and our subjectivation and sociogeny may be in tension (this can often be a good thing, depending on what the world is trying to make us into). Every member of a group is not going to feel or think the same way down to the fine detail, nor is one incapable of going beyond or outside the limits of one's cultures, difficult though it may be. Nor should we assume that some particular phenomenologically unique (structure of) experience cannot be communicated to others to some degree even though those others do not have access to this particular "what it is like"-ness. That we can understand that we do not have a phenomenological structure that others do have is a result of this successful communication. Communication always happens across difference; it does not require phenomenological isomorphism between communicators.

What makes this process more than a repeated subjective commonality? Why must we posit that this is more than what could be achieved through ontogeny or psychic individuation? Is there some reality to the genre or group itself? Sociogeny is not just the production of collective identification. That is, it could not be achieved by merely declaring group membership. A group can be produced well before anyone recognizes its members. Such recognition is a further achievement of a group that can happen only after it is produced. Sociogeny is the production of the intersubjective commonality of groups, of the production of a group identity in the full ontological sense. As we saw with Simondon, any individual must produce a boundary between inside and outside. A group or a genre of human that results from sociogeny does precisely this, and it does so through producing (or having forced on them) a concept of what counts as a group member (or when the group claims to be identical to humanity as such, a concept of what counts as human). The concept of what counts as a member and the related beliefs, mores, norms, etc. it uses to implement that count are what Wynter calls a group's sociogenic code. This code is less a conscious belief of any individual than a sorting algorithm running within the software that is culture and replicating itself within individual minds. We may think of the sociogenic code as an advance on Xunzi's view that what makes humans unique is that we make distinctions through organizing names. Wynter understands how if that is the case, then distinctions make us as well.

A group in this sense is an individual that aims at its own maintenance and reproduction through its members, often at their expense. This is just the same as how a species persists through its members and their reproduction, but no single member is necessary for its continuity. Often an individual's death serves for the species' survival. The same is true of a human and their cells. There is something almost parasitic about groups in Wynter's account. "The coercive nature of the sociogenic code—its imposition being a constitutive feature of sociality—makes it possible for codes to harness the body's drives for its own sociohistorical ends" (Garba 2025, 157). The nationalist volk is a myth of perfect harmony between individuals in an organic unity as a nation. Wynter's sociogeny as a naturalized account of group formation shatters this myth while nevertheless understanding humans as essentially social. There is an irreconcilable tension at the heart of our nature between individual and group preservation. Social meaning mediates and modulates the relation between action and neurological rewards or punishments, bringing group members behavior in line with group preservation. Furthermore, many groups' sociogenic code include a concept of the human, or worse, an equation between what counts as group inclusion and for what counts as human. The social group, in order to maintain itself, organizes material production for what it counts as human (at its most basic, this is a form of kin-recognition), (Wynter 2015, 218). To put it lightly, these counts differ radically from group to group. What matters for the group is that those who it counts as fully human are those whose survival and reproduction preserve the group itself. One can be subjected to or a part of a group whose code does not classify them as human.

This grounding in the sciences and philosophy of mind allows Wynter to develop a radical naturalized phenomenology. Her account then has the resources to explain how humans are heavily shaped by their sociological conditions to the point of producing forms of consciousness and experience unique to those groups yet which are not completely determined by those conditions. Sociogeny is one process among many; that we are shaped by biological, psychic, and perhaps other individuations, and these processes may be in tension, provides the possibility of becoming something beyond what our social (or biological, psychical, or technical) conditions try to make of us. We see in Wynter a more robust version of the claim we have made throughout that it is not by trying to stand outside the forces that shape us that we gain agency over who and what we are, but by becoming active within them, especially in technologically mediated ways.

There is something still idealist in Wynter's account and in how it has been received. In particular, an important part of her work is a genealogy of different conceptions of the human that European cultures have produced and used to subjugate others. This both identifies a deep flaw in our current conception of the human, and it also heralds the possibility of a new one. Her hope is that in using our ability to reshape ourselves, to become something other than we currently are, we develop ways of being different genres of human together without relations of oppression, exploitation, etc. The question then is what allows us to produce a new conception of the human? How does it go from a mere academic exercise or artistic invention to being the sort of concept that operates at the level of sociogeny, as the counting operation of human systems? How to unite the different genres of human under a single conception in a way that is not exploitative, genocidal, or homogenizing? This is the work of what Wynter calls "the ceremony," that is, the act that re-enacts the heresy of humanism to produce a "*new answer* to the question of who-we-are as humans," (Wynter 2015, 192, original emphasis).

The production of a new concept of the human needs an enormous scale of technical mediation to become effective. Here I think the imagery of a ceremony is misleading in that it makes it seem as if a single event led by philosophers and artists could enact this change. Such a feat would instead require concretization through technical memory and generations of ritual maintenance to be successful. In understanding this act as a ceremony, rather than the prolonged ritual practice of summoning a people, and expecting the intellectual work to accomplish what only technical scaffolding can, Wynter's account is ultimately incomplete. This is symptomatic in that Wynter's general conception of what we are as a species is biosmythos, as biological and narrative creatures, (McKittrick et al. 2018), but notably, not as technical. Narrative may be understood as a form of cultural technology, but the materiality of artifacts themselves and their role in this process is not addressed.

Despite their commonalities and the fact that the latter has much to learn from the former, this is ultimately the difference between Wynter and the Simondonian/Stieglerian view of humanity's self-invention.

Relatedly, the danger of not recognizing that we are shaped by sociogenic processes is portrayed as mainly an epistemic danger, though with ethical and political consequences, (Monahan 2024, 273). Insofar as control is capable of hijacking sociogenic processes, epistemic threat is certainly a part of the danger, especially as control involves the modulation of access to (mis/dis-)information. We have all experienced friends or loved ones who have fallen down some months long rabbit hole of misinformation and emerged as someone we no longer understood, no longer shared an epistemic commonality with, no longer knew how to care for. However, we have seen that control threatens more than just our understanding of the world. It damages our very capacities to engage in self-, and group-cultivation. Our narrative nature may be the source of sociogenic codes that shape us, but control does not interrupt our sociogenesis through narrative alone, nor is narrative capable of opposing control's techniques for undermining social individuation. Monahan concludes that becoming active in sociogeny takes place by learning *from* not just *about* others, and in doing so, being open to being unsettled. I agree this would be good, but this is exactly what the exhaustion of subjectivity makes impossible when it undermines heteronomy, as we saw in Chap. 4. Even if the possibility of learning from others remains, diminished though it would be by control, it is an encounter that happens at the level of heteronomy, not transindividuality. Therefore, learning from does not give us insight into how to nurture and maintain social individuation in the face of control. If so, then Wynter's account is useful but incomplete for our purposes. To complete the account, we must ask about the material and technologically mediated production of codes themselves. For this, we turn to Li Zehou.

6.3 Subjectality

While most known in the West for his work in aesthetics, Li Zehou is a profound and wide-ranging philosopher. His work is a unique synthesis of the Confucian and Marxist traditions with Kant. Like our approach to Wynter, we will focus only on his account of the formation of groups, a process he calls subjectality, though this restricted scope is again lamentable. Similar to how Wynter constructs a naturalized phenomenology, Li naturalizes Kantian transcendental aesthetics by grounding it in a Marxist theory of material engagement via social labor that is mediated and abstracted through Confucian ritual. This is his account of subjectality. The term subjectality sometimes refers to a process and to the product of that process; I will mainly use it in the former way. Subjectality is the process of forming a sort of material collective unconscious of groups via sedimentation, of which I take Wynter's sociogenic codes to be a part. This unconscious structure can be taken up as "an object of labor, and thus aesthetically structured as a source of internal freedom," (Garrison 2021, 121). To see how, we must examine the major moments of the process.

Subjectality is the sedimentation of social practice into group connection and an individual's transcendental conditions of experience, (Li 2018, 237). The time scale of this process can be on the order of vast civilizational change, or relatively short-term in times of rapid adoption of new technologies, though generally it is the former. Importantly, subjectality results from practice, not praxis. Practice, in Li's usage is made up of toolmaking and tool-use. In making tools, we develop new ways of mediating our relationship with the natural world. In tool use, especially social tool use, we undertake what Marx called the "humanization of nature," which simultaneously undertakes a naturalization of the human, (Liu 2000, 167). Li critiques Western Marxists, especially the Frankfurt school, for instead embracing praxis, that is, the co-equal status of material and intellectual production. This is not to say that intellectual production, cultural critique, etc. are useless or merely epiphenomenal, but they are downstream of socially engaged material practice. The partial critique of Wynter above is broadly along the lines of this distinction as well.

From practice, two things then occur. First, forms are discovered or invented. In transforming material with tools, we come to notice patterns in nature, we develop styles of movement that are more effective, and we learn how to move in our environment in new ways. The notion of form here is not a transcendent Platonic form, nor even the hylomorphic Aristotelian form. Instead, it is very close to Simondon's notion of implicit form. In working with material to create something, one has a form in mind that they want to impose, but the matter that they are working with already has some implicit forms that must be discerned and incorporated into the crafting process (e.g., the statue already in the marble). The skilled craftsperson is the one who knows how to discern the implicit forms of the material they are working with, choose well, and modify the form they want to produce in response. As such, these forms are not transcendent as they are mediated through human action and tools. They are also not hylomorphic because they are not imposed on unformed matter but negotiated along with a matter suffuse with implicit forms. They result from the interaction of human projects, the objective realities of human social labor and available tools, and the nature of the material transformed by that labor.

Second, through practice (when it goes well), we have a feeling of success. This feeling of success coupled with our feeling or intuition of forms that is then mediated socially that is the source of beauty, (Garrison 2021, 121). On Li's account, our sense of the beautiful is cultivated via the transformation of nature. As we labor and experiment, especially when we do so together, we discover the beauty in things and we invent new beautiful things, (Banka 2022, 122).

> When I talk about "humanized nature," I am referring to the productive labor of making and using tools or the concrete material activity that changes the objective world. I think this is the real origin of beauty... It is not individual's "essential powers" of feelings, consciousness, thoughts, wills, etc., that create beauty; on the contrary, it is the essential powers of the *sociohistorical* practice of the *totality of humankind* that create beauty—Li 1985, 464–5, quoted in Liu 2000, 139, original emphasis

This process transmutes desire into disinterested satisfaction; that is, one learns to appreciate certain forms, first through their coupling with feelings of success, then in their own right. It becomes worth some sacrifice on my part to do things in a beautiful way or preserve a beautiful form.

Our appreciation of beauty leads to ritualization. In ritualization, what was labor bound by natural necessity becomes social and cultural activity. Through aesthetics, we free our movements, our tools, and what we make with them from their necessity in nature and the labor process. Ritual is the preservation of this freedom. It is for these reasons that Li equates beauty with freedom. Moving in rhythm makes certain forms of collective labor more efficient or even possible. By laboring in rhythm, we enjoy the success of the practice and the form of collective action in time. We experience this as beauty. We may then pursue that beauty again by engaging in similar rhythmic movements when not laboring, thereby creating a dance. These practices and rituals shape our bodies, the refinement of our tools, the environment around us, and our cognition (Banka 2022 argues that Li's account of cognition accords with 4E approaches in cognitive science, particularly enactivism, (Banka 2022, Chapter 5)). In turning Kant on his head, Li argues that this is the production both of individual transcendental conditions of experience, and of a sort of material collective unconscious. That is, all of these processes sediment into the cultural conditions that shape us down to our structures of experience. This sedimentation is then structurally similar to Stieglerian technical memory.

Insofar as technology mediates our relationship with nature, it plays a fundamental role in the production of beauty. Furthermore, new technology makes possible not just new forms of labor and social practice, but if we are careful and attentive enough, new forms of beauty, of aesthetic experience more generally, and of cognition and group belonging. It is worth noting that despite the reference to the totality of humankind above, cultural differences are possible and desirable on Li's view. Differences in practice and the forms of tool-use central to a culture will produce different forms of ritual, different aesthetics, different ways of appreciating beauty, etc. These are shareable and can be abstracted from their conditions, but maintaining separate cultures can only be done via a separation of practice and sedimentation.

Li's theory of the social unconscious as resulting from the sedimentation of material engagement sublated into aesthetic forms is a much-needed corrective to our central form of social thinking under control: virality. Ideas spread faster than ever; new forms of thinking, new art, travel the globe in seconds. If such new ideas alone were the motor of social change that so much of twentieth (and sadly, 21st) Century theory took them to be, then we would truly be in a Maoist paradise of continual revolution. Today ideas can spread incredibly fast, but they rarely become new forms of life because they spread free of material engagement. Millions of minds receive them, often even endorse them, and yet they disappear barely leaving a trace in memory let alone in the environment. This decides the question between praxis and practice. The politics of awareness, that anemic form of consciousness raising, fails for being disconnected from practice. It is both rarely generated from or concretized in tools and social tool use. The materiality of Li's account also gives form to the critique of Wynter above; it is not that we do not need a new concept of the human nor to actively engage in altering sociogenic codes; it is that those can only be done or be made efficacious through the production of new tools, new material engagements, and new forms of social practice. A similar issue is at stake in Hui 2024 where a call is made for a planetary thinking which makes possible noodiversity, technical diversity, and epistemic diplomacy, all without an eye to what sorts of practices we need to give rise to this new form of thinking or what practices could concretize and ritualize these forms of diversity. If Li is correct, simply getting the ideas right will not do it. Given the enormous production of new work and the ease of access to millennia of texts in the internet age, there are more brilliant ideas currently available and circulating than at any previous time in human existence. What is more important than ever is concretizing them, building them into technical systems, ritually entraining them into ourselves, and living them together communally.

Let us return to ritual and beauty. Practice, the beauty it produces, and the rituals that preserve beauty while nurturing our sense of it are the conditions of producing, maintaining, and caring for social connection and collective feeling. Here there is a connection with Kant's account of beauty. Kant famously defines beauty as disinterested interest. This is not a contradiction; rather it means that beauty is what I am concerned with (what interests me) but which does not offer me some benefit or gain (disinterest). One who sees a painting in terms of its price tag cannot appreciate its beauty because, while they may care for the painting, they do not have disinterest as they see it as an object of potential material benefit. This may sound like a proto-romantic view of art for art's sake or as valuable in itself, but there is equally a social dimension here. Disinterested interest, that pull towards something that does not benefit me though I feel compelled to care for and maintain it, is what draws us beyond ourselves. Beauty is fundamentally unselfish and it deepens our bounds with others. Relatedly, the artist put in conditions of material dependence on selling their art is alienated from the beauty they produce. Hence the increase of artistic homogeneity alongside global capitalism. As Kant argues, beauty is the property of particular instances and cannot be reduced to a set of criteria or formal list of features. That is, the production of beauty cannot be reduced to an algorithm, and so its production cannot be predicted, forced, or coerced with any reliability. This does not mean that an algorithm cannot produce something beautiful by chance, but that there is no algorithm for reliable producing all or only beautiful things. The production of ever more art that is (literally and figuratively) algorithmic, art that cannot even pretend to the beautiful, is a direct result of the coercive conditions of art under capitalism being at odds with the conditions of the beautiful. Beauty is impossible in conditions of mere survival.

We engage well in our cultivation of groups when we do so for the sake of beauty or in line with the beautiful. Li equates beauty and freedom. For the reasons above, I agree. Furthermore, the process that creates beauty is one that opens new potentialities for human action, and so is a form of becoming active. In producing new forms of beauty alongside new potentialities of action, subjectality also deepens interiority. Like ritual, this is conduct that acts on others' conduct, but in a way that one acts on their conduct such that it increases their freedom and promotes their activity. The cultivation of beauty is power in its most positive, ethical form, and this may be why the Confucians defend the necessity of aesthetics in ritual. It is a form of summoning but not trying to create or force a people into being. If beauty that results from the social ritualization of material practice is freedom, then it is directly opposed to the distributed priests of controls diagnosed in Chap. 2. Here we cultivate spaces of freedom for others, and they cultivate them for us. This is an important design principle for interfaces in general, but especially for augmented reality, as well as a reason to believe in its possibility for radical projects. Such technology can create beautiful forms of mediating collective practice, increasing our freedom in the process. This mutual cultivation of freedom often must be pursued indirectly; I can cultivate the freedom of others through ritual constraint on myself and the trust that others will return this care.

Similarly, if Li's account is correct, it is grounds to critique many of the design choices of digital tools. In the ever-sleeker app- or Apple-ification of design, the inner workings of computers are made less accessible to the average user. In doing so, we are less often afforded the chance to encounter the forms inherent in digital architectures directly. We are denied the ability to engage in practice with these digital tools operating under the hood of frictionless interfaces. The relevant social practices are then restricted to niche groups. Thereby, we neither develop the sense of beauty nor freedom they deserve. This is to say nothing of the individualism and atomization that results from much design of this style.

Let's conclude this section with an extended material example (if one would prefer an example of a digital platform instead, they may consider Wikipedia). Much of the development of a skateboard trick is the experimentation with how bodily movements can affect and correlate with the dispositions of the skateboard, and how it can be used to interact with one's environment in new ways. There are indefinite ways to rotate, move, or position a skateboard, but relative to the motions a human body is capable of and the objects one encounters, a finite number of forms are generated (different tricks, such as a kickflip, nosegrind, 360, etc. are forms in this sense) and explored. This is most easily seen in tricks named after degrees of rotation (180, 360, etc.) where there exist any countable number of possible degree spins, but only multiples of 90 degrees take on any real significance, and so become recognized, repeatable forms. Success is certainly enjoyed when it finally occurs. Once repeated success is entrained, one practices making their movements ever more beautiful, ever more approaching some form produced or discovered in these experiments.

A skateboarder comes to find the beauty in environments that their skills and related technical objects allow them to traverse. Often these are environments others do not recognize the aesthetic value of. For instance, certain drainage ditches become overwhelmingly beautiful. It is common for a group of skaters to adopt some forgotten bit of the urban environment, to care for it, repair it, and modify it. They do this for their own enjoyment, yes, but they also care for it for its own sake. They learn what concrete needs, how to cultivate a territory within an urban ecosystem, and the diplomacy needed to deal with the non-skating locals, law enforcement, etc. They develop skills with new tools. In doing so, new norms and new social codes are developed for the group, yet a space of experimentation and freedom is carved out. Eventually, new people appear, whether experienced skaters, those eager to learn, or to be a part of the micro-culture. This is an example of a successful summoning ritual.

6.4 Conclusion

What is important in these accounts is that there are different genres, different types of humans. If there is something that it is to be human, anything like a human nature (and I think there is, that we are the being that makes and remakes itself through technics), then this general category is produced only through the differentiation processes of its subgroups, not their uniformity or a static list of properties shared by all members of the category. Humanity can be produced only by the always changing differential relations between and within groups and individuals.

Building on our critique from the last chapter, we can see how Aristotle's individualism leads to homogeneity. If self-sufficiency is the criteria by which we rank goods, then there is a single model of virtue: the self-sufficient individual. Hence the paradox of friendship he encounters in the *Nicomachean Ethics*: we would all agree the good life requires friends, but true friendship requires depending on each other. To depend on each other is to fail to be self-sufficient in some way. If we fail to be self-sufficient in some way, and virtue aims at self-sufficiency, then we must be lacking in some virtue in order to have friends who we depend on. Therefore, either we have true friends or complete virtue, but never both. If our goal is to be virtuous, then we aim to be self-sufficient; therefore, if we all aim to be good, then we all aim to be alone (this of course conflicts with Aristotle's definition of human as the social/political animal, but it is a conflict that despite recognizing, he fails to resolve adequately). Whoever lives alone must have some expansive set of skills relevant to all aspects of human flourishing; therefore, Aristotle's view requires a homogeneity of the virtuous. Embracing a strong social constitution of the subject (though not a social reduction) means also rejecting this homogeneity. If we depend on others, then compensatory relations are possible, excesses become virtues in the right relations with others, with the right mediations via artifacts. This means distinct roles can be had requiring different virtues. Friendship here is no mystery.

Combining Deleuze and Li, we may say that we need neither hope nor fear, only to make new tools. At the heart of both our subjectivation and our subjectality is artifice, toolmaking. Yet of course, control's ascendancy is itself the result of the production of new technologies and the social practices they made possible. The question we then turn to in the next chapter is what sort of guidance, what values and heuristics can we use for the production of new tools and their social integration given the context of control?

References

Banka, Rafal. 2022. *Cognition and Practice: Li Zehou's Philosophical Aesthetics*. State University of New York Press.

Deleuze, Gilles, and Felix Guattari. 1987. *A Thousand Plateaus: Capitalism and Schizophrenia, Vol 2*. Translated by Brian Massumi. University of Minnesota Press.

Deleuze, Gilles, and Felix Guattari. 2009. *What Is Philosophy?* Translated by Hugh Tomlinson and Graham Burchell. Verso.

Feenberg, Andrew. 1995. *Alternative Modernity: The Technical Turn in Philosophy and Social Theory*. University of California Press.

Garba, Tapji. 2025. "Sociogeny, Race, and the Theological Genealogy of Economy." In *Words Made Flesh: Sylvia Wynter and Religion*. Fordham University Press. https://muse.jhu.edu/pub/93/edited_volume/chapter/4172194.

Garrison, James. 2021. *Reconsidering the Life of Power: Ritual, Body, and Art in Critical Theory and Chinese Philosophy*. SUNY Press.

Horkheimer, Max, and Theodor W. Adorno. 2002. *Dialectic of Enlightenment*. Gunzelin Schmid Noerr. Translated by Edmund Jephcott. Stanford University Press.

Hui, Yuk. 2024. *Machine and Sovereignty: For a Planetary Thinking*. Univ Of Minnesota Press.

Li, Zehou. 2018. *A New Approach to Kant: A Confucian-Marxist's Viewpoint*. Translated by Jeanne Haizhen Allen and Christopher Ahn. Springer.

Liu, Kang. 2000. *Aesthetics and Marxism: Chinese Aesthetic Marxists and Their Western Contemporaries*. Duke University Press.

McKittrick, Katherine, Frances H. O'Shaughnessy, and Kendall Witaszek. 2018. Rhythm, or On Sylvia Wynter's Science of the Word. *American Quarterly* 70 (4): 867–874.

Mengzi. 2008. *Mengzi: With Selections from Traditional Commentaries*. UK ed. Translated by Bryan W. Van Norden. Hackett Publishing.

Monahan, Michael. 2024. Reason, Race, and the Human Project: Sylvia Wynter, Sociogenesis, and Philosophy in the Americas. In *Philosophizing the Americas*, ed. Jacoby Adeshei Carter and Hernando A. Estévez, 1st ed. Fordham University Press. JSTOR. https://doi.org/10.2307/jj.13049265.14.

Mussett, Shannon M. 2022. *Entropic Philosophy: Chaos, Breakdown, and Creation*. Rowman & Littlefield Publishers.

Simondon, Gilbert. 2017. *On the Mode of Existence of Technical Objects*. Translated by Cecile Malaspina and John Rogove. Univ Of Minnesota Press.

Stiegler, Bernard. 2011. *Technics and Time, 3: Cinematic Time and the Question of Malaise*. Stanford University Press.

Thiel, Peter. 2010. Peter Thielat Libertopia 2010. https://www.youtube.com/watch?v=45o8tQMGtvU.

Wynter, Sylvia. 2001. Towards the Sociogenic Principle: Fanon, Identity, the Puzzle of Conscious Experience, and What It Is like to Be Black. In *National Identities and Socio-Political Changes in Latin America*, ed. F. Mercedes. Durán-Cogan and Antonio Gómez-Moriana. Routledge.

Wynter, Sylvia. 2006. On How We Mistook The Map For The Territory, and Re—Imprisoned Ourselves in Our Unbearable Wrongness of Being, of Désêtre: Black Studies Toward the Human Project. In *Not Only the Master's Tools: African—American Studies in Theory and Practice*, ed. R. Lewis. Gordon and Jane Anna Gordon. Paradigm.

Wynter, Sylvia. 2015. The Ceremony Found: Towards the Autopoetic Turn/Overturn, Its Autonomy of Human Agency and Extraterritoriality of (Self-)Cognition. In *Black Knowledges/Black Struggles: Essays in Critical Epistemology*, ed. Jason R. Ambroise and Sabine Broeck. Liverpool University Press. https://www.jstor.org/stable/j.ctt1gn6bfp.12.

Young, Iris Marion. 1990. City Life and Difference. In *Justice and the Politics of Difference*., REV-Revised. Princeton University Press. https://doi.org/10.2307/j.ctvcm4g4q.15.

Chapter 7
Machinic Phylum

> *"We want neither clean hands nor beautiful souls, neither virtue nor terror. We want superior forms of corruption."*
>
> *—Xenofeminist Manifesto*

Following from Li's account of subjectality, we argued that toolmaking is fundamental to the production of sociality and the ability to actively engage with shaping our social individuation. This raises the question of how to do so well, which brings us to our third process: technical individuation. In asking how to become active in technical individuation, our project is somewhat different from the previous two chapters that sought an account of how to become active in subjectivation and subjectality. It is fairly clear how to become active in technical individuation: make, remake, repurpose, or destroy various artifacts. What is less clear is how to do so well, especially in the age of control. To answer this, I turn to Feenberg's theory of secondary instrumentalization and the ancient Mohists' ethics and counter-siege warfare engineering. This leads to an ethic of democratization and creative resistance.

7.1 Simondon on Technical Individualization

A fundamental debate in philosophy of technology is over whether technology primarily shapes humans or humans shape technology. Instrumentalists argue that artifacts are merely tools, and whatever social, ethical, political, etc. effects they have are not the result of artifacts themselves, but the intentions of those using or building those artifacts. Substantialists, on the other hand, take the position that technology has values and a logic all its own which it imposes on human societies, possibly to the point of technological determinism. Rarely do thinkers take the full extreme of either position. More often, binary positions along a single spectrum like this come

M. J. Ardoline, *Interiority Against Control*, Sophia Studies in Cross-cultural Philosophy of Traditions and Cultures 47,
https://doi.org/10.1007/978-3-032-16245-8_7

to exist only after the question ceases to be a living one (this does not mean anyone stops writing on it). These binaries tend to be retrojections once a solution has been broadly adopted, and they are constructed so the solution can be given the air of common sense, so one can look clever by saying "well, isn't it really a bit of both?" That technology and humans co-constitute each other is now the common-sense opinion, but in being so, its question has ceased to be vital (for the rare, laudable example that bucks this trend, see Roden 2025's defense of a new substantialism). As we saw with the critique of post-phenomenologists, the living question is no longer "if", but "how?" That is, how does this co-constitution take place. For this, we return to Simondon on individuation.

Individuation in general explains how anything becomes an individual; artifacts are no exception. However, they are a distinct case in that unlike physical or biological individuation, human beings (or any being capable of intending to design something) must play a role in the process, and it results in a product radically different than that of psychic and social individuation. Notably, Simondon uses the term "individualization" rather than "individuation" when discussing the creation and evolution of technical objects. The reason for this is that the process that forms artifacts is disanalogous with the processes that form physical, biological, psychic, and social individuals. Artifacts are produced from external forces (usually humans assembling them) rather than spontaneously crystalizing in response to energy gradients. Artifacts also require external maintenance to resist entropy. I will, however, retain the term individuation for technical objects as well. First, I think the disanalogy in terms of crystallization may take the metaphor too far when it is applied beyond physical individuation. That is, I am not convinced that the formation of a psychic interior is more similar to physical crystallization than the process by which a technical object comes to play a role as a condition of its own existence. Second, biological (as well as psychic and social) individuals also required a degree of external support to maintain themselves. A self-maintaining artifact is certainly more foreseeable as a possibility than when Simondon was writing. Third, while the degree of intention involved in humans creating artifacts is unique, it is not different in kind in that humans may affect biological, psychic, and social individuation intentionally when they become active in them. Here I prefer parsimony in that if anything succeeds in being an individual, ontologically speaking, we say it results from a process of individuation. This is true even if individuation processes differ in their details and domains.

The technical object's individuation is a joint process of concretization and adaptation, (Simondon 2017, 58). Concretization is the process by which a technical object goes from abstract to concrete through a series of artifacts. On Simondon's account, a technical object is first abstract because it is a design incarnated in an artifact whose parts do not form a whole in any real sense (they do not have a form of internal resonance). The first version of an artifact is likely to be a series of somewhat connected parts that achieve a function but do so through little more than a form of simple linear causality (like a series of dominoes knocking each other over). Through refinement and experimentation, these parts come to serve more than one purpose in the object's operation and become interdependent, while other parts are removed or rendered unnecessary. For example, in the movement from a gas to a diesel engine, spark plugs are removed and the pressure produced in the cylinder becomes the source of ignition. Perhaps this constrains what other parts can do; perhaps it opens possibilities.[1]

Over this process, the technical object gains an internal resonance as the parts become more pluripotent and interdependent. Note that this does not just happen in a single artifact but over a series of artifacts, where each can be very much unlike their previous versions. What is being individuated as the technical object is not so much any particular artifact as it is the abstract diagram of the thing. That abstract diagram, as it changes, is realized in various artifacts produced according to it, but is not identical to them; it is something more like a species than an individual animal. One also misunderstands this process if they think of it as the ideal form becoming more incarnated, as a sort of technical Platonism. All the artifacts in the series are equally incarnated, and the diagram itself never ceased to be a diagram. Instead, it is the diagram itself that becomes more concrete, in the etymological sense of the word, via a growing together of its parts. The diagram itself changes and mutates.

Simondon here distinguishes technical from biological individuation in that the biological being is always completely concrete. If something is successfully alive, its parts have grown together to the fullest extent possible; if they cease to be together in such a way, the biological being dies. The technical being, on the other hand, never becomes fully concrete. Its parts are always to some degree separable. While Simondon seems to take this to be a difference in kind, I think there may well be cybernetic reasons to take it merely as a matter of degree. That said, Simondon recognizes that at a certain advanced level of concretization, the technical object becomes like the living being in that it can be empirically or inductively studied. One does not learn new facts from a basic blueprint; it is completely a product of the knowledge we already have. However, once a technical object is concrete enough, it becomes complex such that we can learn new things (not simply new to us) by studying it. Here is a second reason the diagram should not be thought of as a Platonic form: the technical object *qua* diagram does not exist independently of those artifacts which express it. Instead, it is a form that can come into being, mutate, or cease to be, depending on its process of individuation.

Perpendicular to the process of concretizing its parts, a technical object is also individuated through adaptation. A technical object, like all objects, is not just parts brought together through an internal resonance. It is also a mediation of relations to other objects. The technical object mediates between other technical objects, the natural environment, and human social systems. In mediating, it is productive of a milieu, of a technical ensemble (a collection of technical and other objects that work together but do not produce an internal resonance) in which it plays a role of maintaining. In this sense, Simondon argues that the technical object becomes (one of)

[1] For a more detailed account of Simondon on technical individuation and the lineage of thinkers it follows from, see Smith 2025.

its own conditions of possibility. That is, it produces and maintains the technical niche within a milieu that both makes it possible and can further facilitate its individuation. It may not reach the level of self-cultivation that a human being can engage in, but the technical object, at a certain level of concretization and adaptation, comes to be a condition of its own continuing individuation. The great historical flow of all technical objects and their individuations is what Deleuze and Guattari call the machinic phylum.

This is then the answer to how artifacts and humans co-constitute each other. Artifacts must be designed and built by humans, but they are not mere expressions of human will. They are expressions of human designs coupled with the various capacities of the cosmos; a given artifact is as much an expression of the capacities of steel, electrons, rubber, gravity, and so on as much as it is of human aims. We may decide a function for a technical object, but its concretization process is not up to us so much as it is a negotiation between us, the parts being concretized, and the physical universe as such.[2] This cascades through to adaptation where the various things brought together via the mediation of the technical object alter each other. Such mediation changes the environment in which we find ourselves. Insofar as the technical object must change, the milieu must change to compensate, including human social systems. And again, at a certain complexity, our artifacts exceed our knowledge, becoming a source of potential new understanding. If so, then artifacts may equally escape our aims and values while never ceasing to put pressure on our own adaptations to them, that is, own individuation. Thereby, we may then be shaped by technical aims and values that we never intended. To assume this is always a negative is to have much too high a faith in human reason or goodness.

With an account of technical individuation in hand, we have a different project from the previous two chapters. It is somewhat obvious how we can become active in the machinic phylum. We make and remake artifacts, we maintain or we discard them, we smash them with hammers until our bosses stop buying them, we assemble them into a *dao* and we care for them, and we recognize our role as pollinators in the great flow of machinic reproduction. What is needed here instead is guidance on how to do so well, and to understand how control is potentially disrupting our attempts to do so. For this, we turn to Feenberg on democratization through secondary instrumentalization as a model of the good. For ways of becoming active in technical individuation when we are excluded from democratic participation in that process, we turn to an analysis of ancient Mohist counter-siege warfare as an ethical model of creative resistance.

[2] This holds just as well for digital media *qua* technology, see Parikka 2015, where it is argued that technology is not just an extension of the human but the earth as well.

7.2 Feenberg on Secondary Instrumentalization

Simondon argues that (presumably Western or industrial capitalist) culture sees the technical object as a particular machine rather than in its becoming. That is, we confuse the artifact for the technical object, the completed actuality for the virtual source of potentiality beneath it. Until we, collectively, recognize that the technical object is open according to its relation to its parts and to other technical objects, technological progress will appear as an extra-human, transhistorical force determining us from the outside, (Simondon 2017, 157). In digital objects, this is further hidden by the ease of interfaces, closed source design, zero trust architecture, etc. The digital case is particularly strange as we know these digital objects are being constantly updated, tweaked, altered, etc., but private ownership hides the work and the choices behind those rolling updates. This gives the whole process a false air of inevitability. But if Simondon is right, then ontologically, humans are neither above nor below machines, but at the same level. There is no such thing as transcendence.

Here I turn to the work of Andrew Feenberg who gives both a set of values to guide this process of collectively becoming active in technical individuation and a schema for the various forms of co-constitution of adaptation between technical and social milieus. I find very little to disagree with Feenberg on these points. Furthermore, the relevance of his project despite being formed prior to the ubiquity of the internet is a testament to the fact that control is not reducible to information technologies even if they are its great facilitators. That said, his schema is somewhat cursory, both in account and in how to employ it. This section then grafts his schema and the values animating it onto our project so far in order to flesh out its content. Where I do find it lacking is in his account of activism and of resistance as a form of engagement with technical individuation. It is unclear, for example, how his account of democratizing resistance can separate the case of AIDS patient activism for more ethical experimental treatments from recent Anti-Vax efforts. Feenberg 1995 provides an excellent range of case studies of activism in technical milieus, but it does not produce an adequate ethics of it. The following section takes up this problem.

Our project here connects with Feenberg's on a few fundamental levels that justify our making use of his schema of primary and secondary instrumentalization. First, he argues that beyond the binary of positive and negative freedom, questions about the design and integration of technology are concerned with the freedom to develop capacities, (Feenberg 2001, 182). This immediately raises questions of deskilling and flourishing. Second, Feenberg embraces Simondon's account of technical individuation, especially seeing concretization as providing an account of technological progress that is not strictly linear or deterministic (Feenberg 2002, 186–7).

Technological development is open to influence by social, political, and economic factors (though not as freely as the instrumentalist would have it). Feenberg lays out a program for a more just (i.e., socialist) form of development: it would require democratization, socialization, and innovation. Greater democratization of society generally, he argues, will influence design and adaptation choices towards giving greater agency to the expected users. The design of information technology analyzed in Chap. 2 lends credence to this in that these artifacts express many of the values of the milieu in which they were created, given that most information technology today is produced within tyrannical despotic regimes (i.e., corporations). We can then also expect, if it is achieved, the concretizing of democratic values into design will reverberate through the institutions and structures these artifacts mediate, (Feenberg 2002, 150). Democratization cannot be understood as mere access or voice, but the ability to play an active role in the processes of technical individuation. This requires increased access to education, especially technical education.

Socialization means simply greater public ownership of the means of production. This has only become more pertinent as machine learning and artificial intelligence proliferate. Take LLM's for example. These models are training on humanity's collective output of digital text. That is, their creation is already socialized, though few of us had any idea of the role we were playing in creating them. Yet, despite the collective creation of LLM's through our (uncompensated) labor, their ownership, design, and profitability (if any is ever found for LLM's) are captured by a few corporations. The only way to oppose this mass enclosure of unwitting free labor is to socialize the ownership and design of these models. I leave out profitability as it is unclear that they should be run for profit at all given our collective role in creating them. If anything, it is among the strongest cases in support of Norbert Weiner's call to re-appropriate the sin of simony (the selling of church or sacred powers for personal enrichment) and apply it to the tech industry (Weiner 1990).

Finally, for the process of innovation, Feenberg means a particular pattern of technological progress that is aimed at overcoming various forms of economic and structural injustice. This model of progress would be one based on the previous two processes, and it would work towards collective emancipation from oppression and amelioration of the natural harms of human existence. Feenberg does not work out the details of this process, but we will return to innovation and provide a more robust account through Mohist ethics below. Before turning to this, however, we must explicate Feenberg's schema of instrumentalization to analyze just how such shaping of technological development is possible, and how to connect this with our accounts of self- and group-cultivation.

Feenberg's schema of instrumentalization separates two simultaneous processes. Primary instrumentalization is the process of separating a bit of the natural-socio-technical fabric, reifying it into a technical object with a function, and refining it. In this sense, primary instrumentalization extends Simondon's account of concretization. Its moments are decontextualization, reduction, automatization, and positioning (Feenberg 2002, 178). In order: to be decontextualized, something must be isolated from its connections, whether that be an engine, a hammer, or the human hand. A schema for the now decontextualized thing is extracted by attending to the qualities or functions of the thing that can be controlled and operationalized (Feenberg calls this reduction to primary qualities). Operations of the schema are autonomized by protecting certain functions from easy or direct interference.

Finally, they are positioned relative to their expected subjects of use both as a technical diagram (how will they be used) and as an expression of power (who controls their use). Here Feenberg recovers what is true in Heidegger's account of *Gestell*, (Feenberg 2001, 188). The technical object requires rending something from the social world, turning a part of nature towards efficient functioning, and instrumentalizing relations between people.

However, efficiency underdetermines the design and use of a technical object, and a technical object cannot be used without being reintegrated into the natural-socio-technical fabric of the world. Remaining decontextualized would mean remaining separate from human practice. Secondary instrumentalization is how that technical object is reintegrated into its role as mediating social and technical milieus. Each moment of separation qua primary instrumentalization has a corresponding moment of reintegration. A decontextualized object must be systematized or reconnected to the natural world and other artifacts. This requires technical changes that allow it to function with other artifacts and to exist within an environment. The aesthetic qualities are returned to an object, and they are given new meaning in social assemblages. The watch does not just tell time; it signals class. Trucks share a function, but one can immediately see which conveys hard work and which is a sign of conspicuous consumption whose driver almost certainly works a job free of manual labor. Feenberg calls this process mediation. What is automatized is so only partially and must be integrated into the routine, roles, and life of its users. In doing so, they shape the sorts of skills and possible lives one may lead. That is, technical objects are integrated into a vocation. Finally, new forms of agency are made available to the users of the object, often in ways in excess of how they are positioned. Feenberg calls this initiative, and through it, one takes up new projects and possibilities. These are the subprocesses of secondary instrumentalization: systemization, mediation, vocation, and initiative.

Here we can use this schema to connect much of what was said about self- and group-cultivation in the previous two chapters to technical objects. This highlights the role that artifacts play in both processes, and how, therefore, by engaging with and shaping technology, we shape ourselves. Because building a *dao* is done not simply by choosing a set of values, but by assembling objects (in the broadest sense, not just artifacts) to produce a flourishing path through possibility space, one's self-cultivation must include systematization and attending to vocation. The production and nurturing of beauty as group-cultivation is facilitated by mediation, by the aestheticization of technical objects and the giving of new social meaning to them. It is through these three processes that we are not just shaped by the norms and values within artifacts but actively reshape them to whatever degree they can be. In doing so, those become the background normative structure of our worlds. And through initiative, we explore what potentials artifacts make available for these processes and our capacity to shape them. It is this last part that we have yet to fully theorize, and it is what turns us to the question of resistance.

In our current world, secondary instrumentalization for the vast majority of technical objects is essentially captured by capitalism, and digital technical objects, by surveillance capitalism. The problem is that the means of secondary instrumentalization, and thereby how most technological infrastructure can be democratized, is in control of private corporations and increasingly undemocratic governments. How, then, to free them? The answer is of course to socialize all of this, but once again, our focus in this work is not making abstract political declarations (fun as they are), but to seek possibilities of flourishing. If these means are not under democratic control, then the possibility of flourishing relative to them is through activism and resistance.

7.3 Mohism and Creative Resistance[3]

In the later chapters of Zuboff 2019, she argues against activists developing resistance strategies. Rather, they should focus on policy and the establishment of new rights. Zuboff laments that artists spend their time creating ways to thwart facial recognition through masks or high contour makeup techniques (Zuboff 2019, 489–91), or that young activists spend their lunchtime sharing strategies to avoid detection (Zuboff 2019, 521). She interprets this as mere mitigation that will only numb us as the forces of surveillance capitalism further entrench themselves in everyday life. Such tactics, she claims, effectively accept the new status quo of ubiquitous surveillance. If so, then these creative forms of resistance are capitulation in disguise, a way of making our lives in surveillance capitalism more bearable without making them more just. Rather than developing counter-technologies and strategies that materially undermine the effectiveness of technologies of oppression, Zuboff claims the better use of everyone's time would be "synthetic declarations that secure the road to a human future," (Zuboff 2019, 489). By this she means demanding new rights such as what she calls the right to the future tense and the right to sanctuary. How such rights are to be established in an age of a global dismantling of democracy is left as an exercise to the reader. I take no issue here with the rights that Zuboff proposes. If anything, the list should be expanded, especially to include the right to repair one's devices, for Simondonian reasons given above. However, I do take issue with the means she critiques and the conversational model of effective politics she assumes.

First though, we must admit that Zuboff's account correctly categorizes some practices. For example, Bonini and Treré's *Algorithms of Resistance* catalogues and analyzes tactics of what they call "everyday resistance", following James C. Scott's work and the distinction between tactics and strategy made by De Certeau (the latter is embraced by Feenberg as well, and so the following critique is part of why his account of resistance needs modification). While Bonini and Treré's work is one of sociology, not moral or political philosophy, they seem optimistic about the tactics they document. A representative example of such tactics is an Uber Eats driver

[3] This and the following sections include material from Ardoline (Forthcoming) though it is revised and truncated to fit this chapter. See the original for an expanded, self-contained version of the argument.

using two cell phones, one to place an order, the other to accept picking up the order, then immediately cancelling the order once they have made the pick-up, thereby getting the food for free. I have no argument against this practice itself, but I do not see reason for optimism as practices like this do fall to Zuboff's critique. Insofar as it does little to challenge the dominant and exploitative business practices of these delivery apps, it is capitulation. This is because the practice could be effectively prevented by changes made to the app design, but such changes would not force some amelioration of the corporation's exploitative practices.[4]

While the title of their work is *Algorithms of Resistance*, almost no attention is paid to the actual use or production of algorithms by activists. We are given few if any examples of activists engaging with technical individuation. Rather the focus is on how people use and appropriate various platforms. This is not a failure of the activists or the "everyday" resisters that Bonini and Treré study, but of the framework of everyday resistance that the authors use. This framework obscures other, potentially more effective resistance strategies. To be everyday is to be a part of one's form of life. Simply put, if resistance is part of one's everydayness, then so are the structures one is resisting. If all that is left is everyday resistance, then Zuboff is right that resistance is capitulation.

Bonini and Treré do take account of some forms of collective action, but these are generally spontaneous collections rather than sustained and long-term. For example, they say collectively gaming an algorithm can be thought of as a digital riot, (Bonini and Treré 2024, 42). The recent move to the app RedNote in response and opposition to the TikTok ban is perhaps an example (Ewe 2025). Yet these spontaneous actions mean little if they do not constitute a successful form of social individuation. Insofar as such spontaneous actions fail to concretize into structural changes that lessen algorithmic exploitation, then Zuboff's criticism holds here as well.

The reduction of resistance to the everyday or the spontaneous and the exclusion of activists from technical individuation misses an entire class of resistance strategies which Ettlinger 2018 calls "productive resistance." Ettlinger identifies the common feature of productive resistance as "the production of new elements of the digital environment that make use of existing elements to produce new truths. The vehicles for such productive power in algorithmic contexts are new material elements, ranging from new software and metadata to new platforms and heuristic tools, as well as immaterial cooperative relations in platform cooperatives and cloud protesting," (Ettlinger 2018, 9). I broadly accept this account of productive resistance, though I will argue below that what matters is not the Foucauldian "production of new truths" but rather what the Mohists will call "benefiting the world."

To defend productive resistance against the capitulation critique, part of what is needed is an ethical principle that does two things: first, it helps distinguish between resistance practices that merely let power flow and those which are potentially

[4]This of course assumes the practice remains an individual one; it is an open question whether it could be scaled up to an effective protest action if coordinated with a large enough group.

disruptive of power structures themselves, and second, it provides a justification for forms of resistance excluded by other common ethical principles without falling into a completely permissive ethical nihilism. To do so, we leave behind the Confucians temporarily to draw from their first great critics: the Mohists, "China's first real social critics and political activists" that we know of (Fraser 2016, 5).[5]

The Mohists were groups which followed the example and teachings of Mozi (Master Mo), a fifth Century B.C. figure who was arguably China's first philosopher,[6] the first great critic of Confucianism, and a legendary paramilitary leader and anti-war activist. While he is less well-known today, Mengzi complained that in his time, the philosophy of Mozi "fill[s] the empire," (Mengzi 2008, 85). The majority of what we know of Mohist doctrine comes from a single text, the *Mozi*, which like most of the texts from the Warring States period was assembled over several centuries. This text claims to preserve the doctrines of Mozi along with dramatized dialogues, and the scientific, logical, and engineering knowledge assembled by Mohists over those centuries. Roughly one fifth of the *Mozi* is a manual for counter-siege practices. While they disappeared from the historical record during the Han Dynasty, the Mohists remain relevant as their disappearance may have been partially a result of their success in that several of their central doctrines were adopted by other schools.[7] However, the ethics of resistance through technological innovation that I defend here is uniquely Mohist.

As the Mohists are less studied than their Confucian and Daoist contemporaries, it is worth giving an overview of their principles, though I will not defend (and in fact suggest we reject) several of their central doctrines. The main doctrines I will defend is their claim that we ought to act to benefit the world and prevent harm from coming to it and that this is done partially through technological innovation. Let us call this the Structural Imperative. I will defend it as a viable principle for engaged, technological resistance. However, having a holistic reconstruction of Mohism is necessary for interpreting the Structural Imperative. At the most general, the Mohist's highest principle is *yi*, generally translated as righteousness, which they oppose to the Confucian priority of *ren*. The Mohists do not reject *ren* outright, but they take *yi* to explain what the humane person would do. Mozi glosses righteousness as *zhèng*, which is translated alternately as "correctness" (Mozi 2010, 263) or "good governance" (Mozi 2020, 100, see 219 for a clarification on the translation debate). In other words, the highest goal of the Mohists is a world with the proper

[5] In some sense, this is still a Confucian exercise as many of the Mohists' rival thinkers adopted and adapted some of their central tenants into their own thought, Mengzi and Xunzi included.

[6] The argument is over whether Kongzi's *Analects* counts as philosophy or not in that its contents are generally assertive and exhortative rather than argumentative and dialectical, whereas the *Mozi* is more the latter. See Fraser 2016. Historically, Laozi was often considered China's first philosopher, though the book attributed to this potentially mythical figure, the *Daodejing*, is now agreed to be dated later than Kongzi and Mozi. The *Daodejing* is also more in the style of the *Analects*.

[7] The *Mozi* itself all but disappeared until it was rediscovered when the Daoist Canons were published in the 1400's. As to why the Daoists preserved the writings of an opposed school, see Goldin 2011.

institutions, governmental structures, and the like. This is righteousness. Like most of their contemporaries, the Mohists do not meaningfully distinguish ethics and politics.

To explicate righteousness *qua* good governance, the Mohists maintain core doctrines known as the Ten Theses, which are generally organized in pairs. The pairs are: Promoting the Worthy and Identifying Upward, which assert the need for a meritocratic hierarchy based on virtue where one has a moral exemplar above them to imitate; Heaven's Intention and Understanding Ghosts, which argue that *Tian* should serve as the model to be imitated at the top of the virtue hierarchy and that ghosts reward the righteous and punish the unrighteous; Moderation in Use and Moderation in Funerals, which critique elaborate state spending and wasteful funerary practices (usually as a part of Confucian rituals), especially when the people are impoverished; Condemning Music and Condemning Fatalism, which critique wasteful public ceremonies and Confucian fatalism about events and character,[8] and finally, Impartial Caring (sometimes translated as "Universal Love") and Condemning Offensive Warfare, which are the most important to our account here.[9]

Impartial Caring demands that one treats all people as of equal moral worth. One has moral obligations to everyone regardless of one's relation or lack thereof to them. This may well be the first defense of universal moral regard in recorded history. It is directly opposed to the Confucian value of filial piety, where one has greater moral obligations to one's family members, and lords and emperors have greater obligation to their subjects but little to no obligation to anyone else. This doctrine is justified by the other theses, By Heaven's Intention, *Tian*[10] accepts sacrifices from all and expects righteousness from all equally, and *Tian* is the model to be imitated (*Tian* offers sacrifices to no one and never errs, so he is atop the hierarchy of virtue), so one ought to imitate *Tian* by treating all others as of equal moral concern. This equality is not inconsistent with but rather scaffolded by their meritocratic views (Promoting the Worthy and Identifying Upwards) in that equal moral regard does not mean equal virtue. Some people are more successful at following *Tian*'s model than others. Those most capable of acting with universal moral regard are to be put in the highest positions of power, with everyone else ordered accordingly. The highest human position of this meritocracy is the emperor, the Son of Heaven, who is explicitly required to imitate *Tian*, as any son should imitate a virtuous father. Similarly, the theses condemning wasteful rituals and ceremonies are justified by Impartial Caring because such wasteful expenditures benefit the wealthy and powerful at the expense of the common people. It is also in the Impartial

[8] The degree to which fatalism was a Confucian doctrine at all is controversial, as the great Confucian texts of the *Analects*, *Mengzi*, *Xunzi*, *Great Learning*, and so on all focus heavily on ethical cultivation.

[9] These are very brief summaries and do not do justice to the richness of the *Mozi*, or Mohist arguments in favor of these theses. For a full reconstruction of Mohist doctrines, see Fraser 2016.

[10] Unlike the account of *Tian* we saw in Chap. 5, the Mohists saw *Tian* as personified to the degree of making choices, and of explicitly punishing and rewarding individuals based on their actions. *Tian*, however, remains immanent on this account.

Caring triads of the *Mozi* that we get the statement of the Structural Imperative: "the task of the benevolent person is surely to diligently seek to promote the benefit of the world and eliminate harm to the world," (Mozi 2020, 56). This statement is generally taken to commit the Mohists to indirect consequentialism.[11]

Finally, Condemning Offensive Warfare does what it says on the tin: offensive warfare is never justified, and should never be undertaken. The reason for this is that offensive warfare always violates impartiality. The *Mozi* offers the following argument: to kill one person is unrighteous, as is to steal from one person because both acts show a partiality for oneself or one's family over the person killed or robbed. Offensive war is killing many people and stealing from many people, which is partiality for one's state over others. Offensive warfare then merely multiplies impartiality and unrighteousness. It is absurd to believe that multiplying unrighteousness can produce righteousness. Therefore, offensive warfare is inherently unrighteous. In terms of consequences, offensive warfare will never cultivate righteousness because the degree of partiality required for such warfare cannot produce impartiality. This is a radical position at the time as it is today. The Confucians justify offensive war if one side can claim to be acting virtuously in doing so (claiming to depose a tyrant)[12]; the Legalist Han Feizi, on the other hand, argued victory was the highest desire, and so any war one could win was justified (though one should aim to not desire unwinnable wars), (Ip 2022, 94–5).[13]

These two theses, Impartial Caring and Condemning Offensive Warfare, are paired together. This pairing is usually understood to be because the case against offensive warfare depends on Impartial Caring. However, I think it is underappreciated by contemporary commentators how important the other direction is: what Condemning Offensive Warfare requires gives content to what it means to benefit

[11] The attentive reader has likely picked up on the fact that the ten theses leave the Mohists with a Euthyphro dilemma: is it *Tian*'s intention that makes righteousness good, or does *Tian* intend it because righteousness is good in itself? There is debate over how and whether the Mohists could resolve this (see Johnson 2011). In that I am rejecting the personified *Tian* of the ten theses, I do not address this issue here. Doing so may well be in line with the historical development of the Mohists, in that a personified *Tian* is a part of the chapters that were written earliest, but the later writings of the *Mozi*, the Mohist Canons, seem to abandon appeal to *Tian*, perhaps a sign that they chose their consequentialism over their divine command theory (Schemmel and Boltz 2023, 65).

[12] See also Lee 2022 on the debate over wars of unification, which Confucians defend to end the chaos of the Warring States period and which the Mohists reject.

[13] Fraser argues that the Mohists have three classifications of warfare: offensive, defensive, and punitive. Offensive warfare is condemned outright. Defensive war is justified if it prevents harm coming to the world. Punitive war, however, is undertaken to punish an unrighteous state. This may sound like the Confucian defense of just war, but the Mohist criteria for a punitive war is so high it is unclear if anyone could meet it. One would need to have Heaven's Intent on their side, and *Tian* would have to somehow choose a state as its instrument of punishment. Mozi gives the analogy that it may be a father's intent to punish his son for misbehaving, but that does not justify the father's neighbor to discipline the child without the father's sanction. In the text, it seems that Mozi gives these criteria to justify the actions of the ancient sage-kings, whose model he argues current rulers should follow, yet makes the criteria so impossible that no living ruler could satisfy it, (Fraser 2016, 203–11).

the world via Impartial Caring. To understand this, we must explicate the Mohist practices of defensive warfare and the role that technological innovation plays in it.

However, before moving on, I want to note again that I am not here defending all ten theses. Rather, I am using them to explicate the meaning of Impartial Caring, Condemning Offensive Warfare, and ultimately, the Structural Imperative, all three of which I do defend. While Understanding Ghosts or Condemning Music help us make sense of these Mohists doctrines (for example, their consequentialism and a form of material economic analysis, respectively), one need not believe in ghosts or stop going to concerts to accept that universal moral regard means rejecting offensive warfare (or to defend the priority of material analysis). Doing so is no different than how contemporary virtue ethicists are not committed to Aristotle's sexism, arguments for natural slaves, requirement that the virtuous person be attractive, or his limiting of ethics to a narrow aristocratic male audience, even though exegesis of Aristotle's texts may require engaging with these views.

7.4 Mohism on Innovation

We cannot specify the Structural Imperative without studying how the Mohists themselves employed this principle. Taken in the abstract, we may misunderstand the Imperative as nearly synonymous with the Utilitarian principle to maximize the greatest good for the greatest number of people. Many scholars treat the Mohists as Utilitarians *avant la lettre*, given their consequentialism, impartiality, etc. I will argue that if we look at Mohist resistance practices, it becomes clear that their imperative is radically distinct from the Utilitarian principle that one ought to do what promotes the greatest happiness for the greatest number of people, and that the Mohist is preferable.[14]

As Pierre Hadot argues, ancient philosophical schools cannot be properly understood unless we take account of the relationship between their doctrines and their practices and understand how both were in service of a particular way of life (Hadot 1995). For example, Epicurean negative hedonism is partially explicated not just through the texts and arguments of Epicurus alone, but also through how Epicureans organized their communal living, withdrew from politics, and made tending their gardens a ritual practice. The practices then help make sense of the doctrine. This is as true of the ancient Chinese context as it is of the ancient Greek. So then, the question is what practices explicate Condemning Offensive Warfare and, subsequently, Impartial Caring and benefiting the world? Despite their anti-war stance, Mozi and his followers were not pacificists. Instead, they became an organized paramilitary

[14] Anglophone scholars have often classed the Mohists as utilitarians. See Hao 2006, Johnson 2011, and Chap. 4 of Godin 2020 for overviews of this debate. Regardless of where one lands on this debate, I take it to be clear that Mozi offers a position distinct from Mill, Bentham, and other Western utilitarians. Whether this means the Mohists are not utilitarians in some more general sense or that we should expand the category of utilitarianism to include them is another question.

force that offered their services in counter-siege warfare to smaller states under threat by larger, aggressive forces. They were renowned for their success. Furthermore, this practice of defensive warfare was completely integrated with their cosmological, scientific, and ethical views, (Yates 1979).

Ip categorizes the Mohists as practitioners of "selective pragmatic nonviolence" in that they saw nonviolent methods as the preferred practical means to their ends, but they did not hold nonviolence as an inviolable principle, (Ip 2022, 25). The Mohists are perhaps the great opposite of the beautiful soul. The beautiful soul is one who professes all the right moral values, such as being anti-war, but then uses these values as an excuse to never get their hands dirty. The Mohists took being anti-war to mean stopping offensive wars in whatever ways were efficacious. That said, the Mohists saw counterviolence of any kind as a last resort. Their preference was to prevent offensive wars from beginning or to end them without violence if possible. Impartial Caring requires them to give equal moral concern to the aggressors, and so if they can be made to turn towards righteousness without being harmed, that is preferable to harming them. However, if these other means fail, then counterviolence is acceptable to protect smaller states and their more vulnerable populations. Impartial Caring does not mean unconditional love in the Christian sense as Mozi specifies that those who cannot be prevented from harming the world by other means may be violently removed from the world, (Ip 2022, 77; Mozi 2010, 626–7).[15]

Nor would Mohists avoid taking on positions of power within large states, rather, their demand for meritocracy instead of aristocracy was precisely aimed at making it possible for lower class artisans like themselves to take on positions of power within states, which Ip 2022 calls trans-grassroots activism. Doing so would then give them some control over the levers of power that made war possible, giving them greater ability to prevent it. Ip's claim that Mohist meritocracy is a form of activism gives practical content that goes some distance in reconciling Mohist hierarchy with Feenberg's democratization. In the context of opposing aristocratic control of state power, a demand for meritocracy, in this case the demand that artisans with practical skills and expert knowledge be included in state decision making, is a form of relative democratization, given the state structure of which they were demanding this change. Following Hadot's hermeneutic advice, the practical changes this produces gives a more progressive interpretation of the doctrine than its discursive reading alone. This is not to claim that Mohist doctrine is completely consistent with democratization, nor all forms of their practice (for example, in their paramilitary organization, the Mohist leader supposedly had the power of life and death over followers). However, had the school not disappeared during the Han period and their trans-grassroots activism been successful, one could easily imagine a history where the doctrine of meritocracy moves towards democracy or where the Mohists underwent a schism into conservative and progressive variants.

To be successful in opposing offensive warfare, the Mohists undertook a generational project of research in science, engineering, and logic. Their scientific research

[15] See also Chong 2018.

focused mainly on mechanics and optics, (see Schemmel and Boltz 2023). In doing so, they contributed enormously to the development of the technology of ancient warfare. Needham and Yates 1995 provide a treasure trove of examples of Mohist technological and tactical advances in defensive warfare, much of which we know only because they were preserved in the *Mozi* itself. Even Mohist research into logic and psychology found application in their innovation of negotiation tactics used to either avoid or to end wars (Ip 2022, 141–5).

What is important to our project here is that the Mohist doctrine against offensive warfare did not justify principled pacificism, solely non-violent means, or merely signing up to defend a besieged city as soldiers; they certainly were not limited to declarations and demands. These actions are not enough to qualify as benefiting the world. Idealists confuse disagreement with resistance, and the Mohists were certainly not idealists. Instead, their doctrine was understood to require becoming counter-siege masters who not only took charge in such situations, but who developed new technologies and tactics, and did so through an enormous undertaking of knowledge production. One must then ask how is it that improving capacities for warfare, developing new forms of killing and new technologies to make it more efficient, could be consistent with benefiting the world by condemning offensive warfare? How is it, to paraphrase Graham, that the Mohists are simultaneously ancient China's most pacifist and most martial philosophical school? (Graham 1978, 4).

By creating new technologies and innovating counter-siege tactics, they make the world better not just by breaking sieges, but by making sieges less likely, by making besieging a city a greater risk. The Mohists do not create technology or tactics that benefit offensive warfare, only defensive warfare. This benefits the world because offensive warfare is unrighteous. Yet, before the Mohist innovations, there is an asymmetry in favor of offensive forces in a siege. It is easier to besiege a city than to fight off a siege. This and the potential rewards gained makes besieging a city compelling even if it is unrighteous. By creating technologies that change that structural asymmetry in favor of defenders, the world is benefited not simply by ending one war, but by making offensive siege warfare less efficacious and thereby less compelling. This is creative resistance; it is resistance that goes beyond the production of new means and expressing new truths to changing the phase space of some human activities such that asymmetries are produced in favor of the good.

The hermeneutical question is now whether there is any support in the *Mozi* for believing that this project of creative resistance and knowledge production was understood by the Mohists themselves as one of innovation to improve the world. One might be incredulous here because innovation is generally taken to be a negative value in the ancient Chinese context (as in most ancient and medieval contexts). However, the Mohists explicitly criticize their contemporaries for praising tradition and denigrating innovation, (Fraser 2016, 228; Yates 1979, 563). In the *Analects*, Kongzi establishes what will become the Confucian reverence of traditional wisdom by declaring "I transmit rather than innovate. I trust in and love the ancient ways," (Kongzi, *Analects*, 7.1, Norden and Ivanhoe 2023, 21, see also Puett 2002, 42). The Mohists reject this reverence as incoherent, favoring instead creation and invention.[16] Consider this passage from the chapter "Condemning the Erudites [Confucians],"

[The Confucians] say 'The gentleman [virtuous person, *junzi*] follows and doesn't initiate.'

> Answer them by saying 'In antiquity, Yì invented the bow, Shù invented armour, Xī Zhòng invented carts, and Artisan Chuí invented boats. That being so, then are today's tanners, armorers, cartwrights, and artisans all gentlemen [*junzi*], Yì, Shù, Xī Zhòng, and Artisan Chuí are all petty people? Moreover, whatever they [the Confucians] follow, someone must have invented it. [If that is so], then what they all follow is the way [dao] of petty people. (Mozi 2020, 138).

In other words, the Confucians defend an account of cultivating virtue in which one follows the rituals and norms of the classical sage-kings, and that virtue is precisely in following tradition rather than making anything new. The Mohists draw an analogy between ethical systems of cultivating virtue (ritual) and inventing technology. They then point out that the truly virtuous one is the inventor. Otherwise, one is in the absurd position of claiming that one becomes virtuous by imitating the unvirtuous (i.e., the petty). If so, then invention in ethics is to be praised just as much as invention in technology. In fact, it is unclear whether this is an analogy or the affirmation that ritual and governance are forms of technology (a position that may have influenced Xunzi on artifice). This makes the same point about innovation argued for above, but in the context of ethical theory. Both the Mohists and the Confucians claim to follow the wisdom of the sage-kings, but the Confucians do not have an explanation for why. The Mohists do. The sage-kings invented forms of governance and self-cultivation that made it more likely that people will become righteous (the triads on Identifying Upwards explain how this process works). In doing so, they benefited the world. And again, the status of the sage-kings for the Mohists is that they are highest exemplars alongside Mozi himself. Therefore, the highest form of benefiting the world, for the Mohists, is to invent that which makes righteousness structurally more likely. I agree.

Finally, we must ask if there is any guide to this process. The answer from the *Mozi* seems to be that in the general sense, *yi* and *ren* provide guides, but in specific situations, benefiting the world may involve violating righteousness and ethical precepts. Mozi criticizes a prince who is offered the throne by a corrupt duke who had the prince's family killed. The prince resists because it would be unrighteous to accept the throne under these circumstances. Mozi asks why, if the prince recognized that the duke was unrighteous, did the prince not accept the position of power, dispose of the duke, and use the throne to re-establish righteousness? (Mozi 2010, 719; see also Ip 2022, 159). In other words, the violation of ethical principles is sometimes required for restoring righteousness, but only in the cases where that violation has the efficacy to do so rather than perpetuate further unrighteousness.

---

[16] See Puett 2002 for a full exploration of the debates around creation and conservation in the Warring States era schools.

Let us then gloss the Structural Imperative as the injunction that one should not simply act according to the right principles, but to act to alter the structures of the world so that the proper principles become both possible and more likely to be actualized (and so acting counter to those principles is reduced). In Foucauldian terms, this amounts to the imperative to exert power well in a way responsive to current and continuing historical injustices.

7.5 Control and Creative Resistance

Are we in a situation that meets the Mohist criteria for productive resistance? Straightforwardly, yes because of the partiality of algorithmic power in the control of tech oligarchs. Technological oppression violates impartial caring as various groups are harmed more than others. "What are these [fundamental harms] produced from? … We must say they are produced from detesting others and injuring others," (Mozi 2020, 56). This detesting and injuring, Mozi calls exclusion, and demands that we "replace exclusion with inclusion" because this will "benefit the world," (Mozi 2020, 56–7). The world is out of order and must be set right; the actions which set the world right are justified if they do so. Whether Mozi would have recognized them or not, capitalist exploitation, sexism, racism, homo- and transphobia, etc. are all forms of exclusion and partiality, or structural injustice, that find their expression in algorithmic power today. Insofar as algorithmic exploitation restricts access to resources by exploited and oppressed groups and makes them more vulnerable to violence, this constitutes a situation we may think of as a cyber-siege or a distributed digital siege. The intensity of the exploitation and oppression of this situation is the degree to which information technologies mediate nearly every facet of modern life. A siege is no longer something that happens only at the level of armies and cities, but also at the level of data sets and dividuals, modulated to target without needing to physically separate and isolate.

An example will make the case. Take the Texas anti-abortion law Senate Bill 8. As a part of criminalizing abortion, this law mandating the creation of a bounty system whereby citizens would be rewarded monetarily up to $10,000 for reporting and successfully suing those who had abortions and the doctors that facilitated them (Bowman 2022). This bounty system was implemented via a government website where these reports could be made. It is a distributed digital siege insofar as it denies certain people access to an important resource, medical care, via a digital platform that deputizes random citizens. Activists responded by developing automated scripts, bots that would flood the website with useless information. These and related actions effectively shut down the vigilante web service (Massie 2021). I take this example to be clearly good and to be representative of the class of creative resistance (and it has only become more effective through advances in generative AI such as LLMs and agentic AI). Thereby, any ethical principle that excludes it would fail to be a good principle for justifying and guiding creative resistance. Such a form of resistance violates principles like those of rational conversation that Habermas still hopes with prevail in digital public spheres in that they rely on misinformation to work (Habermas 2022). Similarly, they run afoul of Kantian injunctions to act in this world as we would in an ideal one as the intentional spreading of disinformation violates the categorical imperative. More generally, Kantian approaches have a difficult time justifying resistance; in the kingdom of ends, what would even need resisting?

Most importantly, however, this example clearly avoids Zuboff's capitulation critique. Insofar as this form of resistance does not merely insulate some people from harm but rather disrupts the tactic of digital bounty hunting itself, it does not constitute an acceptance of the status quo. Yet, it is in line with the Structural Imperative: a technological innovation is deployed to return some degree of impartiality to the world by preventing unrighteousness in the form of oppressive governance from being carried out. The use of disinformation is acceptable here because it does not hamper any structures of good governance.

The Mohists also differ from classical utilitarianism. This is because of two intertwined reasons. First, Mohists are not individualists (Hao 2006). Second, their highest value is righteousness rather than happiness. The difference between acting to benefit the world versus to benefit the greatest number of people is what makes the difference. While both are egalitarians, the utilitarians only quantify over atomized individual human beings, whereas the Mohists look to the world and its structures. The term translated as world is "*tian xia*" which roughly means "all those under heaven" or "under heaven's rule."[17] Though the Mohists differ from their rivals by defending an account of a personified, judgmental god, it would be a mistake to import the Abrahamic view of a transcendent god to the figure of *Tian*. *Tian* is an immanent god whose will is the order of nature. This means that benefiting the world is not simply to benefit individuals, or even humans alone. Rather, it is to look to the structures of the world and to harmonize them with nature to the benefit of all. Recall above that one of the arguments for impartial caring is that it is a structure of *Tian*'s rule that *Tian* accepts sacrifices from all and makes sacrifices to none. Mohist sacrificial practices also expressed this non-individualism: the Mohists rejected individual communication with *Tian*, accepting only group communication with the divine, (Yates 1979, 561). When something is wrong, it disrupts the order of the world. To benefit the world, then, is to do what instantiates the proper structures (such as impartial care) and make action in accord with them more likely. To prevent harm to the world is to protect and maintain those structures. The proper structures are, once again, what the Mohists call righteousness. This, I argue, gives Mohism the ability to attend to structures, systems of order and control, and even ecology in a way that Utilitarianism is doctrinally incapable of, at least without the addition of increasingly unwieldy epicycles.

As Mozi says "the sage, who takes ordering the world to be [their] task, must know what disorder arises from," (*Mozi* 2020, 50). What is good is what brings the

[17] *Tian xia* is not unique to the Mohists, though their account of it differs from their rival schools. It remains an important concept today as well (see for example Zhao 2021).

world to its proper order, the order primarily defined by righteousness and the egalitarian value of impartial care. This is the difference between simply signing up to help fight against a siege and developing the most innovative and powerful counter-siege tactics and technologies of their time. A fighter may do a good action, but this alone has little effect on the order of the world. However, the production and dissemination of a counter-siege manual, the innovation of new tactics, and the invention of uniquely defensive weapons (like the swiveling ballista known as the Mohist crossbow) fundamentally attempt to change the balance of power between the besieging army and the defending city. These actions are good because they benefit the world by instituting the proper order: one where cities are not invaded and besieged because to do so would lead to destruction and death, only now for the invader.

Notably, it is important that these tools do not increase the city's power to engage in offensive warfare. Similarly, the Mohists would supposedly demand oaths not to counter-attack the besieging army when it withdraws in exchange for their services (which we may see as a social adaptation of their technological innovations, a form of secondary instrumentalization). Doing so is more than producing a certain good consequence; it is ordering the world such that good consequences are more likely. While the Mohists agree with Utilitarians that actions are good if they produce benefit, the Mohists give priority to actions that alter unrighteous structural asymmetries. In short, Utilitarians calculate the good; the Mohists alter the calculations. It is this, then, that ought guide the ways we become active in technical individuation.

We said above that a good principle for justifying productive or creative resistance cannot fall into an anything goes nihilism, especially to the point of licensing horrific acts of violence. While the Structural Imperative does justify counter-violence in some situations, it does so only in very specific causes where counter-violence has predictable benefits in line with the proper order and no other means are available to prevent harm to the world. For example, the Structural Imperative does not justify one of the more distressing forms of supposed resistance becoming common in the digital age: stochastic terrorism. One could see this as an innovation of the digital age in that stochastic terrorism largely operates by using various internet platforms to radicalize someone who then commits terroristic acts. This is done not necessarily by targeting any one individual, but by creating an online community wherein it is likely that someone will become radicalized to the point of deciding to undertake such violence. The Structural Imperative may employ counterviolence for the sake of establishing righteousness, but there are strong limits. Stochastic terrorism harms sociality in general as people become less trusting of public space, social connections, and the like. Trust and mutual care are the proper order of the world. In harming these, it also promotes unrighteousness, whatever efficacy it may have. Therefore, even if it were to achieve one's desired consequences, stochastic terrorism is excluded by the Structural Imperative. This imperative, then, is not a trivial defense of all resistance for resistance's sake.

7.6 Conclusion

Zuboff's call for rights at best answers the negative question of what should not be done to us. Within control, we need positive accounts. In lamenting that people spend time developing resistance tactics and strategies, Zuboff is implying that political resistance does not have a place in the good life. I disagree. Any belief that it is not a part of one's self-, group-, and technical cultivation to engage in action that reshapes their sociopolitical situation may as well assume we have reached the end of history. If so many people have become technological reactionaries, it is that they intuit that much technological innovation is currently focused on harming them. What a perverse situation where the increase in humanity's abilities is primarily self-harming. What worse response than merely asking not to be harmed? Perhaps only the techno-pessimist view that sees democratization as a lamentable compromise and would rather undercut our own capacities (see McQuellian 2022). Instead, why not build a thousands words of Damocles? Such a project would certainly be one worth enjoying, and so we must affirm creative resistance as a part of the good life.

Nor should we reduce becoming active in technical individuation to such resistance. Our focus here should not be seen as succumbing to the continental tunnel-vision of resistance, but as a way of giving content to such demands and connecting them to the good life more generally. The production, maintenance, redesign, repurposing, etc., of artifacts is a central part of our flourishing as technologically mediated beings. Against the deskilling forces of control, engaging in technical individuation is inherently reskilling. It is the generation of new possible skills relative to new artifacts as well as gaining some power over technical memory. Democratization is not just the giving of greater voice and education, but greater ability to participate in and initiate such projects. As we saw in Chap. 2, it is partially through skills and habits that we come to feel at ease in a world. If so, then participating in technological individuation to some degree is central to flourishing.

References

Bonini, Tiziano, and Emiliano Trere. 2024. *Algorithms of Resistance: The Everyday Fight against Platform Power*. The MIT Press.

Bowman, Emma. 2022. "As States Ban Abortion, the Texas Bounty Law Offers a Way to Survive Legal Challenges." Shots - Health News. *NPR*, July 11. https://www.npr.org/2022/07/11/1107741175/texas-abortion-bounty-law.

Ettlinger, Nancy. 2018. Algorithmic Affordances for Productive Resistance. *Big Data & Society* 5 (1): 2053951718771399. https://doi.org/10.1177/2053951718771399.

Ewe, Koh. 2025. TikTok Users Flock to Chinese App RedNote before US Ban. *BBC*, January 14. https://www.bbc.com/news/articles/c24757zpqyo.

Feenberg, Andrew. 1995. *Alternative Modernity: The Technical Turn in Philosophy and Social Theory*. University of California Press.

Feenberg, Andrew. 2001. Democratizing Technology: Interests, Codes, Rights. *The Journal of Ethics* 5 (2): 177–195.

Feenberg, Andrew. 2002. *Transforming Technology: A Critical Theory Revisited.* Oxford University Press.

Fraser, Chris. 2016. *The Philosophy of the Mòzǐ: The First Consequentialists*. Reprint edition. New York: Columbia University Press.

Graham, A. C. 1978. *Later Mohist Logic, Ethics, and Science*. The Chinese University of Hong Kong Press.

Habermas, Jürgen. 2022. "Reflections and Hypotheses on a Further Structural Transformation of the Political Public Sphere." *Theory, Culture & Society* 39 (4): 4. https://doi.org/10.1177/02632764221112341.

Hadot, Pierre. 1995. *Philosophy as a Way of Life: Spiritual Exercises from Socrates to Foucault.* Translated by Arnold Davidson. Wiley-Blackwell.

Hao, Changchi. 2006. "Is Mozi a Utilitarian Philosopher?" *Frontiers of Philosophy in China* 1 (3): 382–400.

Ip, Hung-yok. 2022. *Grassroots Activism of Ancient China: Mohism and Nonviolence.* Lexington Books.

Johnson, Daniel M. 2011. "Mozi's Moral Theory: Breaking the Hermeneutical Stalemate." *Philosophy East and West* 61 (2): 347–64.

Massie, Graeme. 2021. "Activists Are Inundating Texas Abortion Law 'Vigilante' Website with Fake Tips." News. *The Independent*, September 3. https://www.the-independent.com/news/world/americas/us-politics/texas-abortion-law-activists-website-b1914035.html.

Mengzi. 2008. *Mengzi: With Selections from Traditional Commentaries*. UK ed. Translated by Bryan W. Van Norden. Hackett Publishing.

Mozi. 2010. *The Mozi: A Complete Translation*. Translated by Ian Johnston. Columbia University Press.

Mozi. 2020. *The Essential Mòzǐ: Ethical, Political, and Dialectical Writings*. Translated by Chris Fraser. Oxford University Press.

Norden, Bryan W. Van, and Philip J. Ivanhoe. 2023. *Readings in Classical Chinese Philosophy*. Third Edition, New. Hackett Publishing Company, Inc.

Parikka, Jussi. 2015. *A Geology of Media*. Univ Of Minnesota Press.

Puett, Michael J. 2002. *The Ambivalence of Creation: Debates Concerning Innovation and Artifice in Early China*. 1st ed. Stanford University Press.

Roden, David. 2025. "The Good, The Bad and the Grimdark: Why Technological Mastery Precludes Collective Self-Mastery." *Technophany, A Journal for Philosophy and Technology* 4 (1): 1–21. https://doi.org/10.54195/technophany.23513.

Schemmel, Matthias, and William G. Boltz. 2023. *Theoretical Knowledge in the Mohist Canon*. Springer.

Simondon, Gilbert. 2017. *On the Mode of Existence of Technical Objects*. Translated by Cecile Malaspina and John Rogove. Univ Of Minnesota Press.

Smith, Daniel W. 2025. Simondon and 'Technological Vitalism': Technics as a Biological Phenomenon. *Substance* 54 (167): 96–119. https://doi.org/10.1353/sub.2025.a969738.

Weiner, Norbert. 1990. *God and Golem, Inc.: A Comment on Certain Points Where Cybernetics Impinges on Religion*. MIT Press.

Yates, Robin D. S. 1979. "The Mohists on Warfare: Technology, Technique, and Justification." *Journal of the American Academy of Religion* 47 (3, Suppl.): 549.

Zhao, Tingyang. 2021. *All under Heaven: The Tianxia System for a Possible World Order*. University of California Press.

Zuboff, Shoshana. 2019. *The Age of Surveillance Capitalism: The Fight for a Human Future at the New Frontier of Power*. Public Affairs.

Chapter 8
Conclusion: Finding Joy in the Forces that Shape Us

> *"The art of living is the art of killing psychology, of creating with oneself and with others unnamed individualities, beings, relations, qualities. If one can't manage to do that in one's life, that life is not worth living."*
> *—Foucault*
> *"Many are those entirely absorbed in militant politics... Rare, very rare, are those who, in order to prepare for the revolution, wish to become worthy of it."*
>
> *—Georges Friedmann*

I imagine each of us as caught up in an enormous, turbulent river. We were born midway through its course, and we will die before we reach the ocean, our bodies dispersing back into water. We cannot escape it. The turbulence is created by the intermingling flows of individuation: physical, biological, psychological, social, technical, and who knows what else. We are in some important sense the river, some of the parts of it that woke up, but that we find anything important or that anything makes sense at all is because we have temporarily managed to separate ourselves from it. We must cherish this temporary condition because in it, we can do what the river itself cannot. The myriad things flow through the river with us, changing as we do. It is through them, through each other, and through at times letting water in and at times forcing it out that we have any ability to direct our course. There is nothing else to do but to survive, to enjoy, and to enjoy others' enjoyment.

The highest good one can aim for is the aesthetic enjoyment of a flourishing life. If we enjoy properly, even our survival becomes a secondary concern. Life is not for anything; there is no achievement to satisfy it. Of the great harms Christianity has visited on the world, perhaps the worst is its giving to life an instrumental goal: to save one's soul and attain paradise. It makes life an individualist striving for something we are not capable of, and it makes what is truly good appear frivolous and

M. J. Ardoline, *Interiority Against Control*, Sophia Studies in Cross-cultural Philosophy of Traditions and Cultures 47,
https://doi.org/10.1007/978-3-032-16245-8_8

selfish. The same is true of nearly all conceptions of the afterlife.[1] Desire and life are as social as they are individual, and so much is required for the aesthetic enjoyment of a life to be possible. To care for others, to be a part of efforts to oppose injustice, to better understand the world, to cultivate oneself, to know, to build, to love. Those unhappy ambitious ones do not know this. Those with all the goods and money but who must either ignore or hate the masses whose immiseration their own wealth is built on do not know this either. Nor do those whose social interactions, from folly or necessity, are first and foremost manipulation for their own ends. One cannot enjoy in such conditions.

We must build a *dao*. We care for ourselves by caring for a path. There, we cultivate beauty so that we may encounter others in their heteronomy and come together in our transindividuality. We maintain the things out of which our path is built so as to nurture the cosmos and our skills; we invent and we tinker to address and correct unjust structures. In doing so, we become an active part of the processes that shape us, and we can make use of the tensions between them, shaping our artifacts and our groups to shape ourselves, and vice versa. This shifts the distribution of power to some often small but meaningful degree, altering how one is acted upon by others and redistributing the singularities where power gives way to violence. We also open spaces and roles we cannot fill ourselves so that others can become active in their own subjectivation, in subjectality, and technical individuation. In so doing, we shape the flow of power towards experimentation, towards greater distribution, and towards the production of new ways of being together. The ineliminable possibility of tragedy aside, how can one who does all this in the company of others fail to flourish and enjoy that flourishing?

A good life does not require a good world. Understanding the good life does not solve political problems; however, a good life is not possible when checking out from political problems. A good life makes us more resilient to those problems, and it gives some content to what solutions should look like. It is what we seek when we give up myths of salvation, of the world restored to some Edenic paradise or of a perfectly just world to come. I often wonder if this aversion to theorizing the good life is partially due to a feeling of guilt on the part of academics, or perhaps it follows from a worry that to claim a good life is possible here and now is also to somehow diminish the enormity of injustice in the world. No guilt is needed and no such implication follows. The joyful academic and the joyful militant are possible. Canguilhem would point out that those thinkers who wrote of their radicality and their resoluteness in the face of a meaningless, unjust world did little in difficult times compared to Cavaillès (though here Fanon is an exception). When asked how he was able to do what he did, Cavaillès supposedly responded "I am a Spinozist." Spinoza equates feeling one's capacity to act increase with joy. If one is a sad militant, it is not because the world is so unjust (though it is); it is because one's capacities to act have atrophied in the face of that injustice. Joy as the collective cultivation

[1] Nirvana is included here as well due to it giving to life a goal beyond life, despite it not being an afterlife in the strictest sense. An exception may be the Homeric afterlife of the pale shades in that it depicts an existence to which mortal life is favorable by comparison.

of capacities, of experimentation and creative resistance, opposes this and makes one less vulnerable to the pernicious effects of power.

Maintaining the possibility of a good life in the face of control requires us to deepen and protect interiority, of our own as well as others. As should be clear from the account in the previous pages, this is in no way a call to turn inward, to become self-focused, or to give up on lofty political goals, up to and including revolution. It is instead two twin claims. First, any belief that a good life must wait until the world is fixed is fundamentally misguided and is no different in result than believing all suffering in this life is redeemed by a better afterlife. Worse, if good lives are impossible under unjust conditions, this may imply that bringing about that imagined better world is worth the sacrifice those suffering today. Second, injustice must be attacked at its conditions as much as resisted in its effects. Waiting for a just world is a result of the repression of technics, of ignoring the technologies that make history and reshape what it is to be human together. Such repression is a central source of leftist academia's technophobia. Social media and messaging apps have facilitated unionizing in situations it was thought impossible. Digital media made the enormous 2020 protests possible, with acts of solidarity across the world. Protest tactics were exchanged and developed from the Arab Spring to Hong Kong, to the US, and beyond. As I write this, Nepal has coordinated in-person action and wildcat democracy through Discord.

In these moments, those hopes of continental thinkers, the spontaneity of the event and the connections of new possible communities, obtained, and yet the better world these were supposed to lead to remains ever to come. What is lacking here, and more so in their conceptualization, is any of the forces that should take these situational upswells and convert them into long-term movements and global institutions, those forces which would engrave their values into technical memory, into the very conditions of the survival of human life moving forward. What is lacking is everything we avoided building due to an obsession with weakness, moral purity, spontaneous organizing, hoping for the messianic and the uncalculable, and any other way of avoiding power. Technology does not have an inherent right-wing bias. It did not undercut the left. Capitalists captured digital technologies, then enclosed, de-democratized, and turned them into engines of exploitation. Fascists learned to use them and to wield their powers of subject and group disruption. They accelerated control while the technophobic elements of the left believed they could simply refuse it by let beings be.

A spontaneity that does not crystallize into a structure dissolves without effect. This is true both at the level of groups and of individual character. In this, there is a deep sense in which virtue ethics is a dangerous strategic choice against control. Being virtuous means being consistent according to some general account of the good. Being consistent means being predictable, and control is based on prediction. However, this is the gambit: control can handle chaos and spontaneous movement; it cannot handle the integrity of meaningful engagement. It can handle dividuals and masses of data, but it cannot handle coordinated groups. It processes information but does not grasp meaning. Control can predict what it can bring to light and categorize. Interiority builds opacity, and meaning has a density that cannot be mined into data without being lost. Control can modify what it can atomize, operationalize, and dividualize. A *dao* and the groups it summons build sociality and semantic density, returning deep meaning to what has been (primarily) instrumentalized.

Virtue ethics is sometimes critiqued for its reliance on a philosophical anthropology. This is taken to be a grounds for rejection because theories of human nature are out of fashion. Philosophical anthropology cannot be done *a priori*. In exploring humans as beings that are individuated biologically, psychically, socially, and technically, I take it that I have provided a partial philosophical anthropology. It is clearly not, however, a theory of a static or immutable human nature. Instead, it is an account where we are heavily mutable, context dependent, creative creatures. The virtues required for a good life in a particular situation are variable over time and context dependent. This puts a high demand on us to understand our world in order to act well within it. Any attempt to formulate what virtues we should foster that ignores the constraints of the various individuation processes that produce us or the other beings that maintain these processes is bound to failure and harm. In part, what I am calling for with becoming active in our subjectivations and subjectalities is to undertake a speculative and experimental philosophical anthropology. In doing so, we must not be committed to any eternal idea of human nature, nor should we be too precious about the prospect of remaining human.

For now, we are technicians of ourselves and our universe. Ancient rites exist across cultures where the practitioners understand themselves as performing the maintenance of the cosmos. This remains our task. Ritual is the maintenance and tuning of social machinery and the experimental harmonization of that machinery with the rest of reality. We fail only insofar as we neglect our technical nature. Much of our spiritualized and supernatural notions in some way result from the repression of our nature, whether it be the Forms, the soul, a volk, or ancestor spirits. Even though it is understood to be metaphorical, one further repression is the idea that the worst excesses of control are the genie that cannot be put back in the bottle. It imagines technology as something we must submit to once we create it, rather than a flow that we are ever-intermingled with in an on-going process of co-constitution. We repress technics *qua* the signs of our mortality by imagining technology to be a completely external, oppressive force. We fool ourselves into believing that if we could just be free from technology, we would no longer suffer. It is by ceasing to repress our technical nature, by embracing our protheses as both signs of mortality and as means of care for our and others' vulnerabilities, that we have killed every god we have ever imagined. We are the creature that invents itself. What then to us is a mere genie?

MIX
Papier aus verantwortungsvollen Quellen
Paper from responsible sources
FSC® C105338

If you have any concerns about our products,
you can contact us on
ProductSafety@springernature.com

In case Publisher is established outside the EU,
the EU authorized representative is:
Springer Nature Customer Service Center GmbH
Europaplatz 3, 69115 Heidelberg, Germany

Printed by Libri Plureos GmbH
in Hamburg, Germany